the clever body

The Clever Body | Gabor Csepregi

Published by the University of Calgary Press
2500 University Drive NW, Calgary, Alberta, Canada T2N 1N4
www.uofcpress.com

This book has been published with the help of a grant from the Canadian Federation for the Humanities and Social Sciences, through the Aid to Scholarly Publications Programme, using funds provided by the Social Sciences and Humanities Research Council of Canada.

We acknowledge the financial support of the Government of Canada, through the Book Publishing Industry Development Program (BPIDP), and the Alberta Foundation for the Arts for our publishing activities. We acknowledge the support of the Canada Council for the Arts for our publishing program.

COMMITTED TO THE DEVELOPMENT OF CULTURE AND THE ARTS

Conseil des Arts du Canada

LIBRARY AND ARCHIVES CANADA CATALOGUING IN PUBLICATION

Csepregi, Gabor
The clever body / Gabor Csepregi.
Includes bibliographical references and index.
ISBN 10: 1-55238-208-7
ISBN 13: 978-1-55238-208-0
1. Body, Human – Philosophy. 2. Mind and body. I. Title.
B105.B64C74 2006 128'.6 C2006-901685-2

Cover design, Mieka West.
Cover photograph, Getty Images.
Internal design & typesetting, Garet Markvoort, zijn digital.

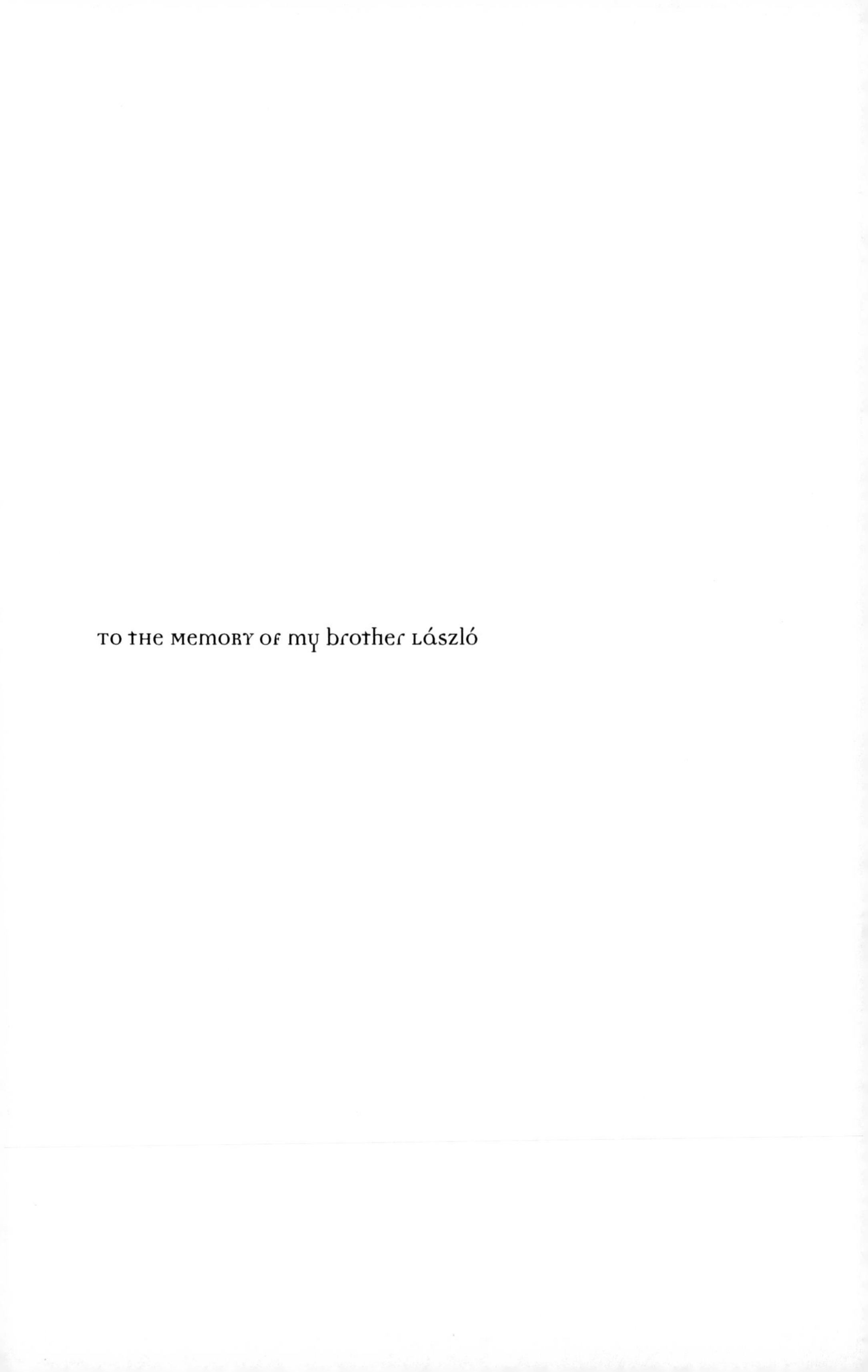

TO THE MEMORY OF my brother László

Acknowledgment | *I wish to thank Mrs. Sandra McDonald for reading the manuscript and making valuable suggestions for its improvement.*

contents

Introduction

Disembodiment | I had long wondered why the philosophy of body was not considered an academic discipline in the same way as the philosophy of mind, language, or art. Yet it is now the focus of academic interest as never before. Eminent scientists undertake, from the most diverse angles, the study of our physiological condition. Numerous analyses deal with the complex relationship between societal and political power and health and cultural practices. Apart from academic circles, the body is in the forefront of our everyday life. Of all ages, and in great numbers, men and women practice some kind of physical activity: they run, ski, swim, or take daily classes of yoga or dance. There is a growing concern – perhaps even an obsession – for the preservation of health, well-being, and youthful appearance. Nowadays, the treatment of emotional and narcissistic disturbances includes the recovery of body awareness, the complete relaxation of musculature, and the refinement of postures and movements. People respond to rock or gospel music by moving their entire body, wearing unusual clothing and costumes, and enjoying an enhanced feeling of togetherness, a sense of *communitas*. Whether one is involved in rock climbing, *t'ai chi*, or the learning of postural improvement, the body is the centre of multiple interests and attentions and is recognized as an essential condition of self-discovery and self-realization.

Some philosophers and sociologists tend to view the "resurgence of the body" with critical suspicion and growing pessimism. They contend that the current exaltation of, and preoccupation with, bodily experiences constitutes a superficial reaction to our long-standing relation of estrangement. In essence, even though the body has become an object of vigorous training and refined care, nothing has changed: "The love-hate relationship with the body colors all more recent culture," argue Horkheimer and Adorno. "The body is scorned and rejected as something inferior, and at the same time desired as something forbidden, objectified, and alienated."[1]

Indeed, when we pay close attention to some aspects of work, education, and health care, we must admit that this judgment is far from being unfounded. Being at the centre of so many interests and activities, the body is vulnerable to marginalization and neglect. Let us consider, for instance, how technological progress affects the bodily dimension of our lives.

The rapid spread of technological devices gives rise to the loss of immediate and intuitive contacts with tangible realities, the growing "abstractness" (*Entsinnlichung*), as Arnold Gehlen called it.[2] Due to the division of labour and the expansion of mechanization and automatization, the sensory contact of workers with the various material realities – stone, iron, or wood – becomes scarce. As the power over our natural environment increases, the bodily interaction with it decreases. The quest for domination and control cannot occur without stepping back from "the personal and immediate involvement in industry and commerce."[3] Therefore, as Albert Borgmann pointed out, the bodily disengagement leads to the gradual degeneration and atrophy of the workers' original skills.[4] Computerization minimizes the sensory aspects of their task. The inability to touch, smell, hear, or intuit the transformation process of various materials produces a sense of loss and vulnerability. The "embodied knowledge" is traded for the "mental involvement," the tactile response to any felt disturbance for the capacity to act upon abstract information. In her illuminating book, Shoshana Zuboff summed up the result of the "reskilling process" in factories: "Absorption, immediacy, and organic responsiveness are superseded by distance, coolness, and remoteness."[5] Where distance is introduced and preserved, the bodily ingenuity can no longer play a significant role. The worker's body has an opportunity to move according to its own natural rhythm, and spontaneously respond to any unforeseen challenge, only if it makes an unmediated contact with some materials.

The growing number of "defensive devices" (Borgmann) puts a considerable distance between the body and the natural surrounding. In addition, they impoverish and flatten the perceptual field. These devices protect not only from temperature and light variations, but also from physical exertion, from the delight of encountering unexpected situations and of overcoming some unforeseen obstacles. Today tourist travel to distant destinations illustrates quite well the ongoing attenuation of bodily commerce with reality. Daniel J. Boorstin remarked that we have lost our ability to travel and gradually became simple tourists.[6] The travellers of previous ages faced the unknown and unfamiliar, dared hardship, kept alive their sense of adventure, and even risked their lives. Modern tourists tend to carefully plan their journey, shy away from all discomforts and risks, and expect "interesting" things to happen to them. Their journey to far off lands is no longer a strenuous and adventurous undertaking. It has become a commodity, a "spectator sport." Aeroplanes, buses, cars, and hotels have formidable insulating effects; they allow "sight-seeing," but no direct contact with indigenous communities. People driving through a natural setting in an air-conditioned car, while listening to their familiar music, remain untouched by the richness and depth of the landscape. They are cut off from the movement of life; they are unable to absorb it. "Such people," observes Borgmann, "have not felt the wind of the mountains, have not smelled the pines, have not heard the red-tailed hawk, have not sensed the slopes in their legs and lungs, have not experienced the cycle of day and night in the wilderness."[7] For many tourists, the perception of a mountain or a deer is little more than a picture seen through the cameras' viewfinder. Since they have no first-hand knowledge of the nature around them, they also fail to engage their bodily strength and discernment; their mediated experience goes together with the idleness of their body.

"The traveler, like the television viewer, experiences the world in narcotic terms; the body moves passively, desensitized in space, to destinations set in a fragmented and discontinuous urban geography."[8] Richard

Sennett's comment refers to our daily urban life. We reach our destinations by car, bus, or train, covering greater distances, without deploying corresponding levels of effort. Robert J. Yudell also holds that "we are increasingly replacing our own body movement with propulsion of the immobilized body. We are replacing motion with 'frozen speed'."[9] In many public places – banks, stores, and libraries – we are even exempted from the task of opening doors with our hands. We encounter more objects than we did fifty years ago but do so at the price of executing a reduced number and variety of movements. If, for some reason, an unexpected mechanical failure occurs, we are challenged by a wide range of almost forgotten actions: we have to climb, bend, jump, and so forth. The disturbance triggers our bodily vigour and skill. A temporary loss of the centralized system of electricity, for example, returns us to an intimate contact with natural fuels such as coal and wood; encountering their resistance, we suddenly rediscover the unused capacities of our hands.

As I ponder the bodily engagement in a technologically shaped city, I learn that runners are now able to use "smart shoes" equipped with a computer chip that adjusts their cushioning level to the runner's size and stride. If this intelligent device is able to "sense" and adapt its shape to the characteristics of the ground, I wonder how such a discharge will affect the sensibility and inventiveness of the legs. Will this relief device make the use and refinement of some bodily capabilities completely obsolete? More questions could be raised when the dream of the "smart home," controlled under one central command, becomes an accessible reality.

To be sure, the swelling abundance of computers, cellular phones and other similar devices produces some very beneficial effects: they create instant connection between people living great distances from each other, promote collaboration of all sorts, and even nurture friendship and love. But the expansion of electronic communication also reduces the number of face-to-face, spontaneous encounters on the streets and generates a web of disembodied forms of communication. It also erodes social skills.

Devices provide people a "hyperintelligence," as Albert Borgmann has shown, but also make them lessen or lose their bodily presence. "The hyperintelligent sensorium, just because it is so acute and wide-ranging, presents the entire world to our eyes and ears and renders the remainder of the human body immobile and irrelevant. The symmetry of world and body falls to the level of a shallow if glamorous world and a hyperinformed yet disembodied person."[10] Disembodiment is not merely a well-informed but also an "unworldly" way to exist. It produces the tendency to ignore the subtle resonances of the body and, as a result, to relate to objects and people with emotional detachment. Furthermore, insensitivity arises from the routine of daily life, the lack of immediate contact with the concrete, and the inability to invest activities and objects with a symbolic content. One's sense of inner emptiness becomes more acute in the presence of an environment that appears impersonal and insubstantial.[11]

R. D. Laing made the distinction between the embodied and the disembodied self. Whereas the former "is fully implicated in bodily desire, and the gratifications and frustrations of the body," the latter considers the body "more as one object among other objects in the world than as the core of the individual's own being."[12] The disembodied self is a performing self in the sense that, preoccupied with appearances, it displays a calculated, controlled, self-conscious behaviour. Not only does it dissimulate moods and desires under a perfectly homogeneous appearance, but it also speaks and acts in an artificial manner. When one's own being continuously becomes the object of critical scrutiny, the capacity for spontaneous actions, intricate rhythmic patterns, and creative mimetic gestures is also paralyzed.

The clever body | Although I am well aware of the numerous factors that promote passivity and disengagement, I am not inclined to say that our bodily capabilities play only an insignificant role in our lives. On the one hand, various human activities – gardening, painting, or dancing – still allow us to rely on our bodily resources and thus feel ourselves, as it were, carried by them. On the other hand, even the refusal or inability to act in accordance with our bodily impulses cannot completely eliminate spontaneous and surprising reactions. An absolute control over the body is just as impossible as is a total integration or a lasting immersed state. Therefore, it would be mistaken to think that, in all our sensori-motor experiences, the disembodied mind proposes and the mindless body disposes.

We may, for instance, consider certain current educational methods and practices pertaining to our embodied life as expressions of resistance and a correction of the noxious effects of our technological civilization. Their purpose is to reject the obsolete conception that cuts off mind from the body and considers the latter as a complicated machine.

Philosophers, psychologists, and students of the anthropological medicine bring to our attention the limits and insufficiencies of a scientific method that ignores all the dynamic and reciprocal relations the body entertains in space and time. One can never understand the living body if one persists in treating it as a self-contained, mechanical structure, unrelated to a wider context. Thanks to its resources and communicative processes, the body continuously transcends its purely physical aspects; it is a dynamic, moving form, an "orientation" (C. A. van Peursen), and an ongoing process of "bodying forth" (Medard Boss).[13]

In the present volume, I would like to show that the body is a mobile structure, endowed with some capabilities that we are able to dent or nurture, but unable to eliminate or create. Just like the heart in the organism, the living body is the source of an irreducible, autonomous, and creative dynamism, indispensable for the multiple relations we entertain with the world.[14] I propose to describe and systematize in some detail the activities that benefit from the body's indwelling wisdom, consisting, above all, of delicate responsiveness and astonishing inventiveness.

I am not concerned with the various physiological systems and automatic processes that secure and preserve an organic stability essential for a healthy existence.[15] The focus here is on our "prereflective body" (van den Berg), on the forms of its attitudes and movements, through which we communicate with things and people. "Prereflective life, that is, life as it is lived in our day-to-day existence, has no knowledge of physiology."[16] Van den Berg's statement applies not only to the acting subject – an organist who plays a fugue with the outmost agility – but also to the observer who analyzes and describes a specific motor experience. F.J.J. Buytendijk is right in thinking that we cannot understand from physiology how an acrobat or a violinist simultaneously executes some complicated movement patterns.[17]

Observing human activities prompted my interest in this study. I was stirred by the obvious fact: "the world is full of movement," in the words of the American dancer Merce Cunningham. Full of intelligent movement. Eminent thinkers have also brought to my attention this kind of capacity, and I am indebted to them for their insights. They have argued that the proper understanding of bodily intelligence is an indispensable condition for the discussion of the basic features of human existence. If we want to throw light on social relations, language, and artistic activities, or examine how we design and inhabit our living space, we cannot ignore how we experience our body. Therefore, the anthropological theories, which disregard the bodily basis of human life, are incomplete and contain serious flaws.

Etienne Gilson, in his lectures on the art of painting, rejects the view of those philosophers who claim that the art resides wholly in the mind and the hands merely execute the orders they receive. He admits that some painters, stressing the all-importance of the part played by the mind in order to gain a better recognition for their profession, are also guilty of generating such a misconception. Experience shows, however, that a painter, though in possession of a representation of the work to be done, relies on the capabilities of a "progressively educated hand." "Man does not think *with* his hands, but the intellect of a painter certainly thinks *in* his hands, so much so that, in moments of manual inspiration, an artist can sometimes let the hand do its job without bothering too much about what it does."[18] If philosophers, declares Gilson, instead of only thinking about art, were required to make a painting, they would "realize how clever the body of an intelligent being actually is."[19]

The Dutch biologist Frederik J. J. Buytendijk is known mostly for his works on pain, play, and movement. In these contributions, as well as in his *Prolegomena to an Anthropological Physiology*, he makes valuable observations pertaining to the body. Beyond the previously accepted distinction between "thing-body" and "lived-body," he emphasizes the difference between the states of being conscious or unconscious of the body. In our non-reflective relation to the world, our body is never a mere apparatus reacting to some stimuli, but an evolving subjectivity responding to meaningful sensory qualities. The responses to the surroundings comprise both activity and passivity, moving and being moved: "Each movement, including the looking, is primarily a pathic moment, a form of self-movement through being moved."[20] Thanks to our already acquired technical dispositions and the awareness of the demand of the actual situation, we know how to perform appropriate actions. Buytendijk speaks of the "available body" and this availability is manifest during the execution of a great variety of movements without conscious control. In addition, our body possesses a remarkable capacity to sense what can and should be done in a situation – it is endowed with a sense of values based on past experiences and open to future possibilities. Because of such an implicit awareness of norms and values, we are able "to think with our hands."[21]

Aldous Huxley's essays on education present a brief but original analysis of our human nature.[22] Consciousness is obviously a central feature of human life. Our conscious self is associated with a certain number of what Huxley calls "merging not-selves." These, for their functioning, do not require attention and guidance. However, our conscious self can affect them in some ways: it can distort or curtail their contribution or, by earnestly abandoning itself to their powers, intensify their influence and effectiveness.

First there is our personal not-self, our habitual way of acting and reacting that is the result of the sum of experiences preserved by our body. Another not-self is our system of autonomously functioning physiological processes. These are in charge, for example, of oxygen supply, digestion, regulation of temperature, or muscular activity. A no less important not-self is our bodily intelligence; it finds and proposes solutions for unique and unforeseeable problems. In moments of inspiration and illumination, we surrender ourselves to a spiritual not-self inhabiting a much wider realm. On rare occasions, the spiritual experience of the "ultimate ground of reality" makes us aware of the "universal not-self."

The aim of education, as Huxley sees it, is not merely the verbal transmission of abstract knowledge – ideas, theories, and information. There is much more to be done than merely sharpening the students' intellectual powers. The body needs as much care and attention as does the mind. "Our business as educators is to discover how human beings can make the best of both worlds – the world of self-conscious, verbalized intelligence and the world of the unconscious intelligences immanent in the mind-body, and always ready, if we give them half a chance, to do what, for the unaided ego, is the impossible."[23] That chance is given to the bodily intelligence when students acquire the art of combining relaxation with effort, "the art of getting out of the way."

For this essay on the clever body, I drew upon numerous other sources as well, particularly upon the contributions of the leading figures of anthropological medicine.[24] The research of these original thinkers is not guided by the traditional dichotomy of mind and body. It is rather concerned with the dynamic and complex correlation between the human subject and the world – a correlation, in which sensing and moving, space and time, reason and emotion, capabilities and opportunities are not rigorously separated.[25] This reciprocal interaction can be healthy or pathological, personal or impersonal, objective or intimate, general or unique, natural or symbolic. Without dropping the demands of objectivity, this approach examines the complex system of Gestalten and also communicative actions in concrete circumstances. It prefers the active and sympathetic participation in lived and global experiences to the detached and analytical investigation of isolated processes. The outcome of such an enquiry is not a "fine theory" but a "plausible insight" that invites the readers to reflect on the characteristics and significance of their own experiences.[26] This book is a modest attempt to achieve this result.

I | AUTONOMY

DYNAMIC STRIVING | All of us notice from time to time – while dancing, skiing, or playing tennis – that our body moves naturally, without conscious control or effort. It not only carries out a given task, but also appropriately responds to unexpected challenges and proposes surprising solutions. Sometimes, as we come to a rest, we ask ourselves: how did we do it? How did we ever come to perform such a movement? We then perceive our living body with a sense of unity and a feeling of harmony. We have the impression of being carried by our body's indwelling energy and competence.

We execute many movements in our everyday life without consciously controlling them. We eat, drink, greet someone, or drive a car with no thought to how we accomplish these actions. In a given situation, we do exactly what appears to be the most appropriate and useful. On these occasions, we do not consider our body as an instrument to be guided and used; it is lived as a silent, dynamic, and reliable support of our undertakings – an autonomous support, moving according to its own rhythm and speed.

Autonomy denotes the ability to act, move on one's own accord. The Greek *automaton* conveys a similar meaning: a being that is the source of its own movement. We may speak, in the wide sense of the term, of autonomy when the movement is prompted by a voluntary decision: I decide to go for a walk and, while initiating and guiding my own movements, encounter no constraint. In a narrower sense, bodily autonomy refers to movements that we accomplish without voluntary decision and conscious attention.

What makes such bodily autonomy possible? Our actions unfold thanks to an ongoing and dynamic striving inhabiting our body. We perceive this forceful striving when, after a more or less long period of immobility, we acutely feel a fundamental need to do something. Children satisfy their inner need to run and play once their class is over. Writers yield to an urge to interrupt their work with short walks. To describe this propensity to move, we may use terms such as drive, desire, interest, or yearning. In all cases, we refer to a primal vital energy that impels us to act or respond. This dynamic striving is present at all levels of our active life: it manifests itself in the satisfaction of our most basic physiological needs as well as in our passion, perception, learning, and quest of knowledge, love, beauty, recognition, or harmony. It permeates the various strata of our being, as well as the most diverse activities that we undertake.[1]

To be sure, many of these activities occur in our everyday space and unfold through a sensory-motor communication with objects. Our primary contact with the world is a "sympathetic understanding," an unmediated grasping of the physiognomic characteristics of objects: we find a street, a car, or a shop pleasant or unpleasant, attractive or unattractive.[2] Our sensation of the immediate appearances elicits a response. In a conversation, we hear more than the meaning of the words, we see more than the face of our interlocutor: we also hear the kindness in the tone of the voice or see a threat in the glance. True, we occasionally tend to detach ourselves from our actual situation and become an objective spectator of an event. We then seek to impose a control over our body by holding in check its propensity to respond instantaneously. Notwithstanding our effort, we are unable to completely eliminate the symbiotic aspect of our experience: we are seized and moved by some characteristic features. Yet, however important such an unmediated communication with objects is, our movements cannot be prompted without the elementary striving of the body. The motor response to a motivating quality does not occur and develop without our body's natural tendency to move.

Play, which begins at a very young age, is doubtless one of the human activities that benefits the most from the body's latent energy. Many playful activities start with an encounter with an object. Because of its manifold possibilities, this object exerts a fascination on the player, elicits a movement, and, once the play is underway, responds to any movement with a counter-movement. The readiness to yield to the object's "invitation" springs from a spontaneous urge to move, a compelling inner impulse to act.[3] We may compare this impulse to the need to take a breath

– a movement, which is neither a reflex reaction nor a voluntary activity. When we hold our breath, we first feel a desire and, later, a strong urge to breathe: we *have* to breathe.

A tangible manifestation of the inner urge is what Buytendijk calls "youthful dynamic."[4] In this context, the concept of youthfulness does not denote a particular period in human life; it does not refer to an age but to a mode of being and moving. One of its important characteristics is the absence of direction: the movements do not follow a strictly prescribed plan and are not tied to specific starting points or goals that could enclose them into a fixed and definite framework. Rhythm is another significant aspect of youthful movement. While very young, as well as throughout our whole life, the rhythmic swinging of our body yields to a delightful play.

The inner striving of the body is one of the elements that make successful theatrical performances possible. Beyond the articulation of the written text, acting principally consists of moving in a particular space, the stage. Actors grimace and gesticulate in order to represent a thought, a feeling, or an image, provide an appropriate illustration for the text, and incarnate a specific role. Eugenio Barba speaks of the "dilated body," a body that becomes the tangible manifestation of thought or feeling. Dilation is not merely the skilful expression of an inner reality. It is also the actor's bodily presence in front of the spectator – a presence consisting of continuous change and growth, sustained by the flow of energies in the body. "The tensions which secretly govern our normal way of being physically present come to the surface in the performer, become visible, unexpectedly."[5] Michael Chekhov further probes this claim and asserts that the body must become animated not only by the energies necessary for the execution of everyday actions, but also by its creative impulses. "The actor's body can be of optimum value to him only when motivated by an increasing flow of artistic impulses; only then can it be more refined, flexible, expressive and, most vital at all, sensitive and responsive to the subtleties which constitute the creative artist's inner life."[6] Contact with the creative impulses endows acting with originality and ingenuity. In absence of this contact, it risks sinking to the level of a non-artistic and shallow copy of some of life's situations. Creativity in acting is rooted in the body, not merely in the bodily striving but, more specifically, in the basic impulse to respond to values and feelings, and to invent original forms.

It is useful, following the fine analysis of Chekhov, to distinguish between striving to do something and striving to achieve something.[7] The former leads to the accomplishment of movements without aiming at a goal. The latter seeks to reach an objective and produce a result. It may consist of impersonating another human being, creating a form out of various materials, or of composing a melody. Having an objective, such as representing some subtle features of a character, or making a painting or a sculpture, does not necessarily mean that the formative activity consciously summons up the bodily striving in order to reach its goal. Gregory Bateson noted that, during a formative activity, artists do not deliberately seek to exploit the body's creative resources. "The artist may have a conscious purpose to sell his picture, even perhaps a conscious purpose to make it. But in the making he must necessarily relax that arrogance in favor of a creative experience in which his conscious mind plays only a small part."[8] Artists, therefore, gratefully welcome the so-called "good moments" during which ideas, solutions, or forms come upon them, and their hands seem to be guided by impulses lodged inside their body.

Both forms of striving elude instrumental control in the sense that we may repress them, hold them in check, or consciously further and orient their dynamism, but we cannot produce them at will. They announce themselves as a continually available energy concealed in our body. Although rather schematic and brief, the foregoing account of some activities makes clear that the body is much more than an object that we are able to hold at a distance and manipulate according to some ideas or wishes. It is, above all, a subject endowed with a general vitality that encompasses all our activities and establishes itself as a fundamental condition of our human existence.[9]

THE CARRYING BODY | Paul Ricoeur considers this involuntary activity of the body, together with the conscious will, as primary anthropological characteristics: "Human existence is like a dialogue with a multiple protean involuntary – motives, resistances, irremediable situations – to which willing responds by choice, effort, or consent. I submit to the body which I guide."[10] Growth or decline, gradual modification of our physical appearance, muscular vigour, or articulate mobility are just some of the involuntary occurrences of our body. In the course of our personal becoming, we undergo several important changes that we have to acknowledge. In a similar manner, moods overwhelm us and we have the impression of being pulled by them. They can be so intense, as in the case of a piercing grief, that sometimes we feel as if they exist independently and control the body. It would be accurate to say that the body, which wants to preserve a peaceful existence, is overpowered by the body.[11]

The observation of these experiences prompted Jürg Zutt to assert that we are truly carried by a certain number of organs, of physiological and psychological functions, and the irreducible fact of "being carried" (*Getragensein*) defines and guides our personal becoming. Being carried somewhere in space and time is one of the original modes of being a body. Paradoxically, we are carried and, at the same time, it is ourselves that carries us. "This being-carried carries us, from the spatial point of view, far in space and, from the temporal point of view, far in time, into the future. We *are* ourselves this carrying that carries us since I am my becoming: I become."[12] In other words, we are delivered to the autonomous vitality of our body since the needs, tendencies, changes, and desires of our carrying body precede and resist our will. When, for instance, we are hungry, we become aware of the modification of our carrying body and the hold that such a state has on us. Likewise, when we are fully immersed in cutting stone or wood, we may note that our own skilful movements are guided by a powerful "creative urge" (*Schaffensdrang*) within the hands.[13]

The body announces itself with its autonomy; without any voluntary decision or planning, our carrying body undergoes some modifications: it becomes hungry, restless, energetic, sad, or tired. Such modifications should not be understood as mere physiological changes. We become hungry in a personal manner, not independently of a specific situation, and in relation to a unity of factors conveying some specific meaning. Unless we are completely exhausted, we become tired when faced with a certain number of tasks we select and pursue. The bodily "not-being-able-to-do" cannot be separated from the subjective "not-wanting-to-do," from our personal response to an invitation or from a request to do something. The body that carries us is not a machine working independently of the world in which we find ourselves with our personal history and projects.

The world presents itself with qualities according to the change that occurs in our carrying body. Therefore, the various bodily modes of being – hunger, fatigue, thirst, or sadness – are not merely inner states, but also ways of finding ourselves in our concrete environment, relating to meaningful things, events, or people, and acting either upon or with them. Restless and agitated, we relate to other automobile drivers in a completely different manner than when we are calm and relaxed. We perceive a house as a welcoming haven if we are suddenly in danger.[14]

An agreeable and convenient manner of experiencing our body is in the state of well-being or fitness. Most often unnoticed, this state is characterized by the pleasure of finding ourselves in good physical and mental condition and having available energy to undertake various tasks. We find the immediate surroundings stimulating and friendly, and tend to relate to them with a sense of unity, integration, and even intimacy. We perceive the road, the field, or the hill as supports of our intentions and responsive to our actions; we view them as means that assist us in our carefully planned or spontaneous initiatives and allow us to reach our objectives. While moving, we may reach our destination with ease and efficiency, or we may ignore the principles of economy of effort and usefulness. We make, then, various detours, jump frolicsomely, or remove and replace things without being able to give account of the functional value of our actions and the ways by which we execute them.[15]

endogenous capabilities | According to Hubertus Tellenbach, the bodily impulses, drives, and urges are endogenous realities.[16] They dwell in the body and move it in a rhythmic manner in order to attain an objective and thus fill a void. Various bodily processes and ways of being, such as being tired, fit, ill, or sleepy – induced by the dynamics of the inner "vital flux" (*Lebensfluss*) – are also endogenous developments. Endogeneity refers to the origin of all these transitory experiences. It is a ground that shapes a manifold of vital processes and events. Some hereditary and permanent elements, such as talent, disposition, typical attitude, body type, characteristic of the intelligence, and dominant temperament, are also grounded in, and emerge from, this "original shaping power" (*ursprünglich prägende Macht*). Beyond some specific aptitudes and constant tendencies, a certain number of bodily capabilities are also rooted in the endogenous sphere. Since it pertains to the life of the individual, this sphere grants to all vital processes, traits, and dispositions a particular unity.

Whatever originates in, and develops from, this patterning force is not at our disposal the same way as, for example, an instrument can be. It is possible to modify the length and rythmicity of our sleep and wakeful state. However, we cannot eliminate their periodicity. We are able to alter our body, but if we do, we succeed only to a certain extent. The process of individual maturation eludes our control. We cannot "will" responses to arise spontaneously from our body. The basic figure of *endogenous*, manifesting itself in our attitudes and movements, does not yield to a conscious manipulation; it resists instrumental domination. When we feel the need for food or rest, or instantaneously overcome an unexpected obstacle, we notice that something happens to us. Tellenbach speaks of the "non-voluntary," "non-disposable" aspects of vital processes, referring to their common feature: the *pathic*. The endogenous aspects of our experiences are not the results of our conscious decision and effort: *we are subjected to them*.

Buytendijk also considers the *endogenous* as a fundamental characteristic of human being. "The 'endon' refers to the hidden ground of the 'authentic' being-able of the person as human, considered as much in his general humanity as in his individual psychophysical existence."[17] This "being able" is understood as both a hidden and a perceptible reality: while conversing, we perceive the act of speaking, but not the gift of speech as such. The human disposition of speaking is both bodily (as the capacity to structure itself in order to produce sounds) and personal (as the capacity to communicate meanings through the body).

Endogenous processes do not occur in an isolated manner, independently of a concrete context: our needs manifest themselves in our daily life; our dispositions and particular competencies unfold "in the full reciprocation" with people and objects.[18] From this follows that we are able to exert some influence on this fundamental interaction. Since our talents and capabilities reveal a significant plasticity, we are able, through appropriate education, to refine and improve them. For Tellenbach, the expression "natural and intersubjective cosmos" refers to actions and material things that give an orientation to the endogenous processes and powers. Thus the *endogenous* is not merely a necessary reality but also a possible and desirable one, in the sense that, by withdrawing our will, we are able to adapt our action to their demand and to enhance their effect.[19] Alluding to Goethe's ideas on the development of the eye, Tellenbach evokes two ways of considering our sensory gifts: we can instruct them, or we can be instructed by them.[20]

Does surrender to the body's autonomous and available dispositions truly offer some beneficial results? Tellenbach speaks of the significant "advantage" of the *endogenous* when he evokes the "real" and "incomprehensible" knowledge that inhabits the organs of the animals and allows them to carry out meaningful actions in the absence of experience and reflection.[21] He believes, however, that "the instruction that a human being can receive primarily from his organs is relatively limited, though not altogether absent."[22] (An infant is instructed by his organ when, for example, he starts to play and experiment with sounds and movements.) The knowledge that allows the animal to adapt itself to the environment can be found in the human body as well. However, according to Tellenbach, such a knowledge plays a less-significant role in the formation of human behaviour than it does in the development of the animals' movement and sensory perception. Humans need, above all, verbal instruction and social interaction.

True, before acting promptly and inventively while addressing a challenge, we must first learn most of the movement patterns. Whether we want to drive a car or ride a bicycle, we must represent and live a particular movement as a global form, a structure in which certain elements receive more emphasis than others. The visual control of these dominant elements must gradually give way to their "understanding." To understand a movement is to grasp and co-ordinate its various elements and, through repeated practice, feel that the form is adequate to deal with the environmental conditions. The adequacy required for cycling is obviously different from that of swimming. All the exercises that we carry out tend to promote a feeling of correctness and adequacy. "If this occurs," writes Buytendijk, "a melody of movement resonates in us and moves us like a dancer."[23] Without the control over our body, and familiarity with a skill, the execution of every single movement would require renewed efforts of assimilation, monitoring, and regulation. As Claude Bruaire noted, through learning, the "formless resources" (*informe énergétique*) of the body become "human" and "available" and we have the movement at our "habitual disposal" (*disposition habituelle*), ready to be used at any moment.[24] Thus, we come to acquire, sometimes not without some toil, a great variety of motor patterns: we learn to walk, jump, swim, throw a ball, drive a vehicle, or play on the piano.

Once a motor structure is assimilated and understood, and the natural dynamism of the body is brought into action, the movements follow each other harmoniously and the necessary adjustments or variations happen by themselves. The body exhibits both its own organic powers and its already acquired versatile technical understanding. Its natural spontaneity has become truly human. The endogenous knowledge of our body announces itself to a greater extent than what is recognized by Tellenbach. The "spontaneous involuntary" (*involontaire spontané*) of our body allows us not only to respond successfully to the requirements of a situation, but also to invent all sorts of new movements. Drawing their energy from the body's natural dynamic striving, spontaneity, together with other capabilities, is an endogenous resource, offering guidance to a great number of actions, from the most elementary to the most unusual.

John Blacking's observation summarizes the train of thought of the present chapter: "Human behaviour and action are extensions of capabilities that are already in the body, and the forms and content of these extensions are generated by patterns of interaction between bodies in the context of different social and physical environments."[25] These capabilities are of central concern for phenomenological anthropology. In the pages that follow, I discuss them in more detail and highlight their significance.

2 | sensibility

The pathic aspect | As we carry out our manifold daily tasks, communicate with people, and move around in our familiar surroundings, we are exposed to a great variety of impressions, to colours, sounds, odours, or tactile qualities. When we enter into a public place – a shop or a restaurant – the strong smell or the loud music literally envelops us and elicits a bodily reaction. As the word "impression" indicates, the sensory qualities impress upon us, affect us. Inside an office building, we may feel ill at ease, or, while conversing, we may be struck by some changes in our partner's facial or vocal expression.

The quality of the building or the inner disposition of the person facing us are lived rather than consciously known and represented. "In entering an apartment," says Maurice Merleau-Ponty, "we can perceive the character (*esprit*) of those who live there without being capable of justifying this impression by an enumeration of remarkable details, and certainly well before having noted the color of the furniture."[1] Indeed, there are many situations in our daily life wherein our reaction to events, spaces, gestures, and words occurs in absence of an explicit conceptual understanding. We relate to objects or to people with an implicit or tacit consciousness: our body knows much more than we are able to explain by words.[2]

We may consider all of these experiences as *pathic* in the sense that they are preconceptual and involve a bodily response. Pathic is the characteristic feature of communication itself: it is a transforming relationship to a situation that personally affects us in some way.[3] When we aimlessly wander in a large hall and hear a gentle, familiar voice – with its particular pitch and colour – even without exactly knowing who is speaking or what is being said, we still cannot relate to the sound in a detached manner. Our body is seized, moved by the quality of a voice. From the induced feeling-tones follows a valuation of the sensory impression. If we are fatigued or cold and a fog presses upon us, we experience these sensations with a greater intensity and this bodily state prompts a particular relation to whatever we perceive; things are viewed as more frightening.[4] When darkness falls and everything becomes quiet around us, we are "touched" by the spell of obscurity. As we enjoy our own bodily stillness and well-being, we become sensitive to the mystery of our lived space. We assign qualities and meanings to our environment according to our own momentary disposition and attitude.

In his luminous analysis of the human sensory experiences, Erwin W. Straus made a distinction between the pathic and gnostic aspects of our relation to the world.[5] By pathic, Straus means an immediate, sensually vivid communication with tones, colours, odours, and tactile materials. Gnostic is the distant and neutral awareness of the constant properties of things. In the pathic, the *how* is grasped, in the gnostic, the *what* is apprehended. In the pathic sphere, we are taken by the momentary impressions and symbolic qualities; in the gnostic perspective, we are directed towards the determinable and objective features.

Since the spatial and temporal characteristics of our communication with objects unavoidably emerge together, there is a corresponding contrast between what we feel and how we move. In the pathic, we focus on what we do in the present and enjoy an unmediated union between our milieu and ourselves. Our movement has no specific starting point or direction and takes place in a space without a system of fixed valences. In the gnostic, we direct ourselves actively to the past and future, and the space around us is articulated through direction, distance, measure, and stability. Here, our movement is tied to a purpose and related to a "historical space."

It would be erroneous to strictly separate the pathic and the gnostic as two alternating aspects or moments of a global experience. The pathic does not appear when the gnostic fades; it does not belong only to particular objects or specific human actions. The pathic pertains to the characteristics of an activity's immediate experience and also to the reciprocal communication we have with things. For example, while making music, the pathic moment may receive a stronger emphasis, but, if it does, this occurs without the disappearance of the gnostic dimension.

Whether the pathic or the gnostic factor becomes dominant, the communication brings into play a particular bodily experience: "where we speak of the gnostic or pathic moments, we are definitely comprehending the experience of the live body (*Leib*) in relation to its surroundings or to the world."[6] How can we characterize the living body being "pathically" exposed to sensory impressions? A central element of the pre-conceptual and unmediated bond is, as I mentioned earlier, the experience of being affected. Something takes possession of us; our body is seized by a quality and delivered to its influence. The decisive factor here is the body's ability to echo vivid and penetrating impressions, to resonate to the appeal of meaningful events. The words "echo" and "resonance" refer to the receptivity and responsiveness of the body, to its attunement to the outside world.

Following the fine analyses of Buytendijk and Jean Ladrière, who both find their inspiration in the thought of Martin Heidegger, we could evoke here the concept of *Stimmung*, the mood, the experience of being-affected, -tuned, -disposed of the body.[7] This expression signifies the body's passivity, its receptivity, its capability to be influenced by outside impressions. At the same time, various valuations, meanings, and modes of behaviour arise out of the mood. Therefore, we find in the *Stimmung* the "paradoxical unity" (Ladrière) of passivity and activity. The body and the concrete milieu reciprocally define each other: the body is moved while it virtually or concretely moves itself and assigns meanings.

In his essay on the body's receptivity to living spaces, Ladrière, widely relying on the views of Michel Henry on affectivity, contends that our bodily responsiveness does not occur only on the level of our intentional-pathic relationship to the world.[8] Our feelings – joy, sadness, contentment, distress – are constituted thanks to the fundamental capacity of our body to affect itself and to feel itself. This self-feeling does not lead to any theoretical representation; rather, it is immediately lived and indubitably apprehended within the context of our personal life. Ladrière admits that the affectivity, as such, cannot be understood solely as the body's relation to itself. It surely connects us with events, allowing us to resonate to a stream of qualities, to be touched and moved by them. He argues, however, that our affective attunement to the other is made possible by the "primal responsiveness" (*réceptivité originaire*) of our body, by its indubitable and effective receptivity towards itself. Beyond its dynamic interaction with things and people, our body is eminently receptive; it has the capacity to be affected by itself and this auto-affection, the immediate passivity towards itself, constitutes the original reality of the living body and makes possible its openness to the world: "We are affectivity, in as much as our being is essentially passive. But this passivity is the body itself, we are thus affectivity in as much as we are body."[9] If external impressions are able to induce in us affective dispositions, this is ultimately due to the capacity of our lived body to affect itself, feel itself, relate to itself without any mediation and distance.

delicacy of the body | To more adequately describe the body's affective communication, I would like to introduce the notion of sensibility. It refers to our elementary pathic responsiveness to appearances, changes, and challenges, the body's constant and unavoidable exposure to impressions. As sensible beings, we are subjected to all sorts of influences from external events and developments. "Because of our sensibility," says Louis Lavelle, "we cannot stand aloof from the world without; through our sensibility, the world acquires a sort of consubstantiality with us; our body is bound to it by mysterious fibres, so that no one of these can be touched without our whole being being affected."[10] The bond between our body and the everyday realities allows us to exist; at the same time, however, it may bring an end to our existence. We are at the mercy of random happenings. Similarly to Ladrière, Lavelle claims that sensibility is not only a primary form of contact with objects and people, but also an essential sort of self-awareness, an immediate and indubitable revelation of our own existence. Without sensibility we could hardly recognize our body as belonging to us and, to a certain extent, to everything that affects us. It is because of its constitutive vulnerability that we are eminently conscious of our body: we become conscious of what can be affected, perturbed, or even lost.[11]

C. S. Lewis remarks, in his *Studies in Words*, that "sensibility always means a more than ordinary degree of responsiveness or reaction."[12] Indeed, being a mere vulnerability, a pathic exposure to a particular outside influence does not exhaust this particular bodily capacity. It is through our sensibility that we become aware of changes, contrasts, differences, deviations, even if our discrimination remains vague and lacks precision. Lavelle speaks of the "delicacy of the body, which reacts to the subtlest and most remote happenings outside it, enabling it to distinguish their finest differences."[13] We meet someone and, if we are attentive enough, can sense immediately that something is not right, something has happened, or she is just not the same as before. Or if we encounter two brothers, we are able to detect instantly, sometimes with an extreme accuracy, qualities that distinguish one from the other. We sense small differences in their manner of greeting us or in their facial expressions, yet we are unable to tell what exactly prompted this recognition. Like a very delicate and sensitive seismograph, our body is able, from a very early age, to register and respond to variations, nuances, and shades in attitude and behaviour of people around us.[14]

Our body's discriminative sensibility relates to the surroundings in accordance with our activities and interests. When, for instance, we deal with important practical issues, we tend to ignore superfluous or distracting sensory impressions. We pick out zones of interest and become indifferent to irrelevant information. The focus on specific tasks or objects prompts our sensibility to select aspects or areas and pay closer attention to them. Straus illustrates this tacit modification with the following example.[15] Entering a crowded hall, we are assailed by a loud and distracting noise. But as soon as we meet someone and engage ourselves in conversation, the situation changes: we are no longer overwhelmed by the enveloping noise. As we become attentive to the words and gestures of our partner, the steady and disturbing din of the crowd fades away. The same experience occurs when, driving in heavy traffic, we turn on our car radio or use our cellular phone: both the visual and auditory impressions become less insistent. Our bodily sensibility is now bound to a restricted sphere of interest. If, on the other hand, we fail to exhibit an interest in specific occurrences, we find ourselves exposed to a wide variety of visual and auditory information. In this connection, Straus contends that an extreme sensibility to noise, exhibited by sick persons, is due to a disturbed communication with the world and not to an altered functioning of the acoustic nerves.

This brings us to another important aspect of our bodily sensibility: its relationship to movement. Whether focusing on particular objects or relating to a broader range of situations without a specific concern, our sensibility, involved or disinterested, is bound to movement. Touching manifests this relationship in a striking fashion. When we touch an object, the impression obviously originates from the movement of our hand. If our hand stops moving, the tactile impression remains constant and eventually ceases to affect us. It is the deployment of our movement that makes possible the affective resonance to an object. On the other hand, our movement itself is the result of the tactile contact: the hand's exploring movement is guided by some tactile qualities – hardness, softness, smoothness, or roughness – and the affective tones induced by these qualities. The nature and the intensity of the movement depend, to a great extent, on the affective attunement to the object. The motor competence of some basketball or water polo players are surely enhanced by the playful relationship that their hand entertains with the ball because, being a round, simple, and perfect sphere, it naturally exerts an attraction. Repeatedly catching and throwing a ball evokes in the player a particular satisfaction: in Buytendijk's words, the "feeling of co-existence without any resistance."[16] The same feeling could be elicited by the caress of an infant's head or any other similar round and smooth object.

A caress, of course, can have various motives and meanings: its affectionate gentleness may be perceived under the aspects of reassurance and comfort, or exploration and discovery. But whatever is expressed, the rhythmic contact of the hand with the body of the beloved person is prompted by the affective component of the tactile impression. Following Jean Nogué's suggestion, we may compare the caress to music-making: just as the flow of sounds provides an impulse to the subtle and delicate movements of the violinist's fingers, the feeling-tones stemming from tactile contacts induce the approaching and withdrawing motions of the hand.[17]

The execution of tactile movements is not a "groundless process" (Straus); we introduce variations of rhythm, speed, and form into our movement patterns because we stand in relation to our milieu and establish a subtle communication with its significant elements. As sensible beings, we experience things, events, and people that speak directly to our body and thus induce various types of motor responses.[18]

Different types of sensory impressions exert distinct affective resonances. The encounter with an object may involve the simultaneous collaboration of two or more senses. In a concert hall, we see the pianist playing a piece and, at the same time, hear the melody. If we close our eyes, we are aware only of the sounds, but not of the moving fingers. While watching the same concert on television, we might do the opposite: we turn off the sound and observe merely the pianist's manual dexterity. The switching from seeing to hearing or from hearing to seeing is not just a matter of focusing or shifting our sensory attention. When we suddenly "turn off" a sense, our experience is not merely less vivid and intense. The change is qualitative: we establish an altogether different kind of relation to the concert itself. If, while watching a frightening movie or a sporting event, all sound is suddenly cut off, the scene ceases to take hold of us. A converse transformation occurs when the "gate" of an additional sense – tasting or touching – is opened. As Straus has shown, in passing from the tactile sphere to the visual or from the visual to the audible, or the reverse way, we experience a significant change in the manner things, people, or events affect us. Objectively, these realities remain the same; what is altered is the mode of stimulating our sensibility.[19] "At an emotional level," observes Anthony Storr, "there is something 'deeper' about hearing than seeing; and something about hearing other people which fosters human relationships even more than seeing them."[20] Lovers know that a gentle touch of the hand of the other creates a more intimate and personal form of relationship between them than the mere exchange of words or glances. In passing from seeing to touching, their relation reaches a different level: the distance yields to an immediate reciprocity, the possibility of participation to a claim for exclusivity.

wider spectrum of the senses | Our body's sensibility and its achievement cannot be explained only by the impressions received by the five major sensory organs. Recent research has called our attention to the significant role that other sensory systems play in our experiences and relationships.[21] When we travel on an aeroplane or stand close to a loud speaker, we are able to detect vibrations of different frequencies. We receive vibratory sensations when we establish tactile contacts with objects, especially with machines using combustible substances. If so many people today are infatuated with a motorcycle, a ski-doo, or a sea-doo, their ardour is caused, in part, by the vibratory sensations they feel at the point of contact between their body and the vehicle. They are thrilled just as much by the changing magnitudes of vibratory impressions as by the high speed and power of the conveyance they drive.

By touching an instrument, deaf persons are exposed to rhythmic vibratory impressions and some even claim to enjoy music. David Katz published a thorough study of a "deaf music enthusiast" who, sitting at a distance from the orchestra, could not only feel with delight the different sound waves streaming through his body but also distinguish the specific character of a musical composition. Katz believes that the powerful effect of organ music on some of us can be explained by the strong vibratory impressions that, in addition to the majestic sounds, emanate from the instrument.[22]

Our body is also responsive to thermal impressions, even if our clothing shelters us to a certain degree. As we go out from our house to the street, we immediately become aware of changes in the temperature. Whether we realize it or not, we always undergo a specific thermic stimulation: the thermoreception of our body is not under our conscious control. Exposure to high temperatures during a hot summer day forces us to admit, sometimes not without irritation, how much our body is bound to, and dependent upon, a specific environment. The heat pursues us, we are unable to hold it at a distance and get away from it. Jean Nogué rightly notes that variations in temperature deeply affect us and alter both our experience of the lived space and our bodily attitude and conduct.[23]

We establish the meaning of rooms, houses, or streets on the basis of our internal state, natural affinities, and actual temperature conditions. In warm weather, the space around us appears friendly and inspires confidence; on a cool day, things seem to resist our intentions. Scorching heat is connected to the feeling of burden, the bitter cold to that of hostility. Warmth has a natural kinship with life and it is for this reason, perhaps, that our body seeks it the most: we enjoy swimming in warm water or lying in a warm bed. Warmth gives to our living spaces the character of intimacy, familiarity: it arouses the desire to abolish distance, eliminate all forms of separation. The warmth of hearth creates a more relaxed and intimate atmosphere and, consequently, fosters a better communication among individuals.

Bodily sensibility is usually rooted in specific sensory systems that function independently from, or in synchrony with, each other.

However, we may respond to impressions or signals without being able to identify the sensory system inducing our response. Our experience is unified, global, and occurs without the awareness of a particular mode of sensing. René A. Spitz speaks of "coenesthetic communication," which is based on a "total sensing system" of the body. Coenesthetic responses are not localized; they are extensive and involve a pervasive sensibility. "The sensorium plays a minimal role in coenesthetic reception; instead, perception takes place on the level of deep sensibility and in terms of totalities, in an all-or-none fashion. Responses to coenesthetic reception also are totality responses, e.g., visceral responses."[24] Infants are exposed, in the first months of life, to changes of equilibrium, tension, vibration, rhythm, contact, time duration, and tone, arising from their immediate contacts. They register impressions not through separate sensory channels but by the coenesthetic organization of their body.

In a similar way, Daniel N. Stern draws our attention to the "infant's formidable capacities to distil and organize the abstract, global qualities of experience."[25] These qualities can be captured in terms of intensity, movement, pleasure, urgency, or shape. Elicited by the vital processes of the infant's life, they give rise to a global and elementary interpretation of events and people, and may later lead to creative expressive activities.

Spitz stated that such an undifferentiated and non-verbal appraisal of a situation or a person, disappears from the life of many adults. The coenesthetic communication tends to diminish in the course of our development and is replaced by a diacritic, conscious perceptual experience. "Our deeper sensations do not reach our awareness, do not become meaningful to us, we ignore and repress their messages."[26] Those, however, who "deviate somehow from the average Western man" and seek to transgress the rational mode of thinking – musicians, dancers, acrobats, painters, poets, and others – are still able to retain and reinforce the functioning of their coenesthetic organization. During their creative moments, they are able to experience a form or an event in terms of "deep sensibility stimuli." They remain attentive to the slightest changes around them, detect coenesthetic signals, and respond to perceptual phenomena with the totality of their body.[27]

In his description of the coenesthetic perception, Spitz refers to the sensibility of the infant in relation to his nursing mother. There are other forms of interpersonal communications in which the coenesthetic reception may play an important role. When members of a string quartet make music together, they follow the notes on the music sheets and hear the tones of the instruments. These sensations give to their experience a diacritic character. However, the successful synchronization of the different musical events requires that each member plays his own part and, at the same time, takes into account the performance of the others. As Alfred Schutz has pointed out, a co-performance calls for the awareness of the measurable time and the sharing of the lived temporalities, in which his own part, and that of the others, unfold.[28] The violinist hears and anticipates the cellist's interpretation while he takes into account the similar hearing and anticipating, by the cellist, of his own play. This complex process of communication and "tuning-in relationship" is not possible without sharing space with the other members of the quartet. The synchronization and reciprocal experience of the inner time of the co-performers demand the immediate sensory perception of the other's bodily expressions. Beyond the perception of a series of tones assigned by the composer, the participants are oriented towards the postures, facial expressions, and gestures of the others. Their body's deep sensibility perceives and interprets all these messages as meaningful signals. If, as Anthony Storr suggests, playing in a string quartet is an exhilarating experience, the enjoyment is no doubt due to the richness and vividness of the non-verbal, non-directed communication between the musicians.[29] Their delight is provoked not only by the production of ordered sounds but also by the sheer experience of togetherness, the "mutual tuning-in relationship" (Schutz), and this is based on the ongoing process of subtle, hardly perceptible, but profoundly meaningful coenesthetic exchanges.

style and atmosphere | Style is the personal and distinctive use of variables (rhythmic structure, key centres, tone colours, tempo deviation, and so on) that make up a musical composition. The recognition of an individual character, the result of many years of practice and experience, calls for the capacity to perceive a unity beneath the diversity of forms, the *Stilgefühl* – as the sensibility for integration and stability is aptly called in German. Thanks to this feeling and before any reflective analysis, we are able to identify the author of a particular artistic work. After hearing a few notes of a sonata, or casting a glance at a painting, we say that this must be by Schubert or Raphael, even though we are unable to articulate our judgment in words – a judgment that is primarily subjective and prone to errors.[30]

It is surely so that artistic sensibility presupposes the repeated encounter with forms. Once experiences have been accumulated, we accomplish an immediate discrimination and sense the characteristic manner of composing or portraying – the acoustical or visual fingerprints. The ability to recognize the personal style and identity of a composer or a painter is based on our talent to echo some invariable elements. Copies or imitations can be discovered, or doubts about the authenticity of an artwork prudently advanced, because our body is able to feel, with equal confidence, the absence of a personality or its presence.

Our feeling for style is operative outside an aesthetic context. Thus, we encounter individuals who express their personality through their gestures, attitudes, and ways of moving. "A style," according to Merleau-Ponty, "is a certain manner of dealing with situations, which I identify or understand in an individual or in a writer, by taking over that manner myself in a sort of imitative way, even though I may be quite unable to define it; and in any case a definition, correct though it may be, never provides an exact equivalent, and is never of interest to any but those who have already had the actual experience."[31] Towns and cities also have their own distinctive character. Merleau-Ponty explains that the style of a city consists in a distinctive and singular figure, a fundamental and concrete structure, an affective essence that is discovered at the very first encounter with certain material realities. We do not have to roam through all the streets, squares, parks, and combine these impressions into a totality in order to get acquainted with a style. Although partial and inexplicit, one glance at some streets and houses makes us aware of a definite style, and although further corrections and subsequent unfolding may make it more complete and articulate, they do not touch its indelible feature. In the following passage, Merleau-Ponty relates how the streets nearby the railroad station in Paris gave him an access to a distinct style:

Paris for me is not an object of many facets, a collection of perceptions, nor is it the law governing all these perceptions. Just as a person gives evidence of the same emotional essence in his gestures with his hands, in his way of walking and in the sound of his voice, each express perception occurring in my journey through Paris – the cafés, people's faces, the poplars along the quays, the bends of the Seine – stands out against the city's whole being, and merely confirms that there is a certain style or a certain significance which Paris possesses. And when I arrived there for the first time, the first roads that I saw as I left the station were, like the first words spoken by a stranger, simply manifestations of a still ambiguous essence, but one already unlike any other. Just as we do not see the eyes of a familiar face, but simply its look and its expression, so we perceive hardly any object. There is present a latent significance, diffused throughout the landscape or the city, which we find in something specific and self-evident which we feel no need to define.[32]

A city presents a material configuration just as much as a human face displays its basic components. Thus, each city has a pattern or a system composed mostly of paths and places where the various elements have a specific purpose, alternate in a rhythmic manner, and communicate with each other. Such a physical organization can be represented, studied, and described through a definite set of abstractions. But a city also possesses a "latent meaning," an "emotional essence" that requires a concrete, bodily apprehension. We become aware of this essence while weaving in and out of side streets, touching the doors and trees, and seeing the houses and stores. As we move through an alley or a park, remaining alert for subtle and complex perceptual cues, we detect the underlying affective quality of our specific surroundings. Tony Hiss notes that people concerned with the preservation of historical buildings describe their own experience of the individuality of a place in terms of "character," "essential spirit," "quality of life there," or "livability, genius, flavor, feeling, ambience, essence, resonance, presence, aura, harmony, grace, charm, or seemliness."[33] Their interest in the protection of historical districts depends upon the affective dimension of their experience, alluded to by these expressions and terms.

Jean Ladrière believes that, in the emotional essence of the city, two basic feelings, joy and sadness, are united in an incomparable and peculiar way. They represent the fundamental aspects of our human condition and destiny, namely the imminence of decay and disappearance, and the promise of perfection and completion. Each city conveys sadness and joy, despondency and exaltation, emptiness and fullness, and, whether we are its visitors or inhabitants, we sense, with our body, their incomparable unity.

Kent C. Bloomer and Charles W. Moore also consider the body as a source of an affective reaction to architectural forms.[34] They relate the environmental meaning to the unconscious and changing image we have of our body and to the various values, feelings, and "psychophysical coordinates" arising from it. Reciprocally, our experience of the physical coordinates of houses develops and modifies our body image. In addition, the body image is extended to visible objects through our haptic sense. We relate to buildings as if we were touching them and, through our tactile imagination, we project on them some of our internal states. Qualities such as heavy, stable, protective, or centred depend for their meaning on the haptic sensations we have preserved in our body.

The city of our childhood leaves a decisive mark on us. We remain forever attached to the first impressions of a familiar yet mysterious space because of the strong emotional bonds created during the early part of our life. Every ensuing approach to an unfamiliar place, as well as every preference and feeling, is tied to our childhood experiences. The meanings we give to new houses and streets embody a decisive reference to the "city of our heart."[35]

The styles that we consider to be the underlying essences of cities and towns are, in fact, atmospheres. These are affective qualities that we detect in our immediate or wider surroundings. Because they touch and move us, in the deepest senses of these terms, atmospheres are, in the words of Gernot Böhme, "stirring emotional powers" (*ergreifende Gefühlsmächte*).[36] We may resist these powers or yield to their compelling influence, but we cannot eliminate them. Wherever we are, in a small room or in the middle of the ocean, we are constantly exposed to a particular atmosphere. Even though we do not always notice it, the contact with an atmosphere is just as much a fundamental feature of our existence as are consciousness and language.

The nature and functions of atmospheric emanations have been illustrated and analyzed with remarkable subtlety by Hubertus Tellenbach.[37] He speaks of the "atmospheric mode of being human" and considers it as one the most important fields of study for philosophical anthropology.[38] Our sense of smell gives us a primary access to an atmosphere. Hospitals, schools, churches, apartments, all give off a particular odour, endowing the whole spatial structure with a certain tonality. Odours, like sounds, detach themselves from their sources, permeate the lived space, and induce a reaction. There is, however, a difference in the ways by which we are affected by odour and sound. Whereas the former encompasses us rather gently, discretely, without inducing a shock or a significant resonance, the latter – sound – exerts a more compelling influence and elicits a more marked response.[39]

Newborn babies achieve a primary mode of contact with their mothers through their olfactory and gustatory senses. They sense not only the scent of a perfume and the taste of the breast, but also an emotional essence, namely the specific atmospheric tone of their mother. "There is," writes Tellenbach, "in nearly all sensory experiences, a surplus which remains inexplicit."[40] To detect a particular atmospheric quality means to reach beyond the factual, the objectively given: to hear, beyond the sound, the timbre of a voice, and to see, beyond the shape, the glimmer of a colour. Thus we are able to grasp, sometimes with great accuracy, the inner state and character of a person who is speaking or gesticulating. We "hear through" the voice or "see through" the movement, to use Nicolaï Hartmann's expressions.[41] Our first impression of a man or a woman occurs by virtue of such an immediate experience of a distinctive atmospheric quality.[42]

Indeed, a particular atmospheric nimbus permeates human beings and endows their movements, gestures, and words with a certain tonality. The personal atmosphere reminds us of the phenomenon of expression: a glance, a vibration of the voice, a gesture of the hand discloses a "breath," a "halo," or a "fine cloud" that constitutes, for Eugène Minkowski, the "spiritual aspect of a personality."[43] We all have encountered strong personalities who exerted on us a distinctively vivid impression: they seemed to radiate energy, wisdom, and conviction. They have inspired us to bring forth our best. Here is, in a sentence, the description of the powerful effect of a charismatic musician: "Liszt was a guru figure, an enormously attractive personality, and while you were in his magnetic presence, as more than one student testified, you played the piano better than you dreamed possible."[44] In some cases, such a personal magnetism could also become a dangerous gift, tempting its possessor to stifle our true individuality and self-expression. A less powerful personal atmosphere pervades the lived space around every encountered person. We sense a particular presence or aura and, with it, a certain tonality – joy, vitality, sincerity, or sadness – that, like perfume, gradually infiltrates the whole surroundings. Children are keenly responsive to the atmosphere created, consciously or unconsciously, by their parents. As J. Rudert remarked, the parental atmospheric radiation is a "kind of spiritual food" that children need for healthy growth.[45] Their personality is considerably shaped by the atmosphere they "breath in" at home. For the same reasons, students either resent or enjoy their learning experience in the classroom. An interpersonal atmosphere, where trust, confidence, and kindness are felt, is needed to make possible their healthy growth. In general, an atmosphere permeates every sector of our life-world, and influences, to a greater or lesser degree, the characteristics and outcomes of human activities.

Specific objects may also be charged with emotional accents and hence disclose an atmospheric quality. We may find them frightening, strange, pleasant, agreeable, nice, or hateful. Besides their objective, categorical appearance, which is relevant for our practical conducts, they may be endowed with some physiognomic characteristics. The world of children is fraught with a physiognomic structure: they see the flowers, clouds, or furniture as animated and affectively toned realities that speak and respond to them in some way. Even adults perceive familiar objects, with their dynamic and expressive qualities, as pertaining to specific feelings or events. Minkowski tells us that the cherished practical objects on our desk do not merely symbolize or recall a segment of our life: we do not simply supplement our perception by memory images. Our past is truly present in them and animates them. Thus we perceive a "breath of life" in a book or a pen just as much as we detect a "holy atmosphere" in a forest or a mountain. If the theft of some cherished objects deeply affects us, it is due to the destruction of an important part of our being. In his remarkable novel *An Innocent Millionaire,* Stephen Vizinczey reflected on this vital bond between people and their possessions and considered it as the cornerstone of our sense of reality. "Things embody something of the years that drift away and evaporate like smoke. Possessions are proof, concrete evidence of all that has disappeared; to rob a man of what he has is to rob him of his past, to tell him that he didn't live, that he only dreamed his life."[46]

In a similar way, Buytendijk studied the expressive and dynamic structure of forms.[47] An affective quality is embodied in both living and nonliving things: sadness, aggressiveness, or gracefulness may be perceived in a tree, a bird, or a cloth left on a table. Artists are particularly sensitive to these expressive qualities. They not only listen to the "suggestions" of various materials but also argue with them. To their hands, a piece of stone or wood is not an indifferent, inert matter; it speaks to them and calls for an answer. The "reactions" and "objections" of their medium often modify their initial conception. Many unexpected solutions and discrepancies between the envisaged shape and the final realization are the result of the artist's attentive dialogue with the artistic medium itself.[48]

Such a questioning communication with the physiognomic structure of "living things" does not occur automatically. Children's activity is not always guided by the expressive quality of objects. Nor do artists, while eating, look for the "hidden meaning" of a tomato or an apple. They come to display sensibility towards the "language of things" when they temporarily suspend their practical intentions and view objects with an attitude of sympathetic receptivity. In other words, things "speak" to them when a matter-of-fact and practical approach yields to a more flexible and playful contact.[49]

Our understanding of the language of forms brings into play a reflective, gnostic and unreflective, pathic moment. On the one hand, the form is apprehended as an objective reality, independent from the perceiver, and subsumed under a definite idea or concept. On the other hand, the various affective meanings are grasped immediately and without any conscious representation. The body senses the significance of the form and shapes the appropriate motor response. Thus, during a mountain hike, we encounter forms with their expressive qualities – the soft murmur of the brook, the gloom of the forest, or the gay light of the refuge – which are not ideas, but meaningful realities grasped by our body and inducing a particular motor behaviour.[50]

Our atmospheric sensibility cannot be reduced to a one-directional attunement. Sounds, odours, and colours certainly come to us, press upon us, and resonate in us, but, just as we like to approach flowers and smell their pleasant perfumes, so, in the same manner, we like to focus actively on some impressions, and reinforce their effect. Minkowski uses the French *aspirer* when he refers to the active aspect of our atmospheric experience.[51] With all our being, we are able to detect and inhale a particular sensory or moral atmosphere without, of course, taking, literally, a larger quantity of air into our lungs. It is worth remarking that, for Minkowski, the act of *aspirer* – as well as that of seeing, tasting, or touching – is not only a distinctive mode of sensory contact with an object but, above all, a fundamental way of being in the world. Thanks to a phenomenological approach, we are truly able to grasp the function and significance of this vital and dynamic category of human life.

I have already alluded to the various responses to atmospheres: streaming traffic on a busy street induces a different reaction than a peaceful meadow. In general, a natural environment seems to arouse in us a higher degree of sensory and atmospheric alertness than does the ambient noise of a large city. "A quiet place that offers no threat seems to invite people to redistribute their attention, and any number of subtle perceptual cues can then come into play."[52] All those who consciously create or modify our immediate surroundings cannot ignore the possible effect of an atmospheric quality on our behaviour and mood. Architects, city planners, landscape designers, artistic managers, or party hosts must be well aware of the correlation between an atmosphere and the way we respond, act, and feel. Performers and lecturers must learn to correctly apprehend and modify a prevailing atmosphere. They all should know that atmospheres can exercise a significant power over human sensibility which, as Paul Valéry observed, is not only a "faculty of sensing," but also a "mode of reaction," "mode of transmission."[53]

The responses to atmospheres are not consistent. Music tends to unite the listener and singer and thus creates a "community of consonance" (Straus). But a particular song, played as a background to conversation, instead of inducing a vivacious participation and an experience of intimacy, can sometimes produce adverse effects. The character and intensity of our responsiveness depend on a great variety of factors, such as taste, culture, living habits, as well as our will, awareness, desire, and momentary mood. These determine our reaction to music, whether we display an attitude of enthusiastic acceptance or that of strained resistance.

The link between the hearing of tones and the bodily response of the listener has been repeatedly pointed out. "Even a seemingly motionless subject generates muscular activity while listening to music," remarks Nils Lennart Wallin.[54] Anthony Storr expresses the same idea: "If we find that a piece of music *moves* us, we mean that it arouses us, it affects us physically. Bodily involvement always implies some kind of movement, whether it be tensing muscles, swaying, nodding in time, weeping, or vocalizing."[55] According to John Blacking, creative listening and music-making not only engage our own body but also produce a heightened awareness of our emotions.

Although the musical conventions with which it can be expressed are part of a cultural system, like the syntax of a language, participation in performance (by listening carefully as well as by actually playing) can involve the body's sensorimotor system in such a way that people's responses to the music are felt as an expression of the very ground of their being and an intrinsic part of their human nature. In literally being moved, both internally and externally, by participation in musical performance, they can become more aware of the human body and its repertory of sensations and emotions.[56]

As already mentioned, it is the pathic character of the sounds that elicits the bodily response. Music presses itself on our body and compels us to move. When we thoroughly relish a musical recital and come to understand it in our bones, we can hardly refrain from moving our hands or some other part of our body. The particular form of our movement depends on the characteristics and organization of the music and the sense of form acquired in a specific cultural environment. If our body responds differently to march music than to a waltz, this is due to the particular harmonic, melodic, and rhythmic patterns of the musical movement as well as to our learned susceptibility to forms.[57] Music's enticing effect is not merely the consequence of the listener's ability to "hear and think music muscularly." Culturally learned habits equally shape the response to musical forms. Hearing and understanding music involve both bodily sensibility and the intelligent recognition of some patterns in sound sequences.

The perception of style, atmosphere, and physiognomic characteristics is based on an intimate and sympathetic experiencing of the world. It is our bodily sensibility that allows us to open ourselves to these realities in an immediate and nonreflective manner. This contact involves both an awareness of a meaning and a dynamic adjustment to a situation. The nature and variety of motor responses arising from the body are the subject of the following chapters.

3 | Spontaneity

The forming body | In a broad sense, spontaneity denotes the basic dynamism or motor intentionality dwelling in our body and announcing itself in the execution of movements. Whenever we stand up and walk or run – when we do anything – we rely on the body's indeterminate energy, on its "natural spontaneity."[1] Strictly speaking, movements performed in absence of constraint, on our own initiative, but without voluntary decision, we may call spontaneous. And in as much as we carry out habitual actions with no conscious co-ordination of the various motor segments, they also may be considered spontaneous. For example, we do this when, without thinking, we ride our bicycle, effortlessly flexing our arms or legs, rotating our wrists and neck, and shifting our weight. Without thinking does not mean without focus: while moving spontaneously, we may be very aware of the characteristics of the road, such as surface, width, level of steepness, or sharpness of curves. What is lacking is reflective control and analysis of the movement. If, instead of "thinking aside" and relying on our body's ability to organize the motor segments into a coherent whole, we try to analyze the movement structure, we risk becoming stiff, awkward, or strained, and, as a result, committing mistakes.

Our movements unfold as a continuous and self-renewing dialogue between our body and the surrounding world. The latter is a complex and dynamic reality; it presents a great number of objects with ceaselessly changing aspects and meanings. The source of this diversity resides, in part, in the qualities and characteristic features of our actions and feelings. Objects look different when we are agitated or calm, when we move or are at rest, when we stand up or lie down. In addition to their meanings, they exhibit a wealth of material characteristics. We are able to perceive and deal with such abundance through a wide range of flexible and variable movements. There is a correlation, as Arnold Gehlen aptly remarked, between the "enormous potential diversity" and "wealth of possible combinations" of human movements and the great number of everyday situations in which we find ourselves and interact with things.[2] Driving a car, for instance, calls for an ongoing motor adjustment to the requirements and challenges of the actual circumstances. It is impossible to steer a vehicle with rigid, stereotyped movements. When we sail a small boat, our body responds adroitly to changing external conditions: we modify the form of our movement and our body position according to the strength and direction of the wind or the height of the waves. Rock climbers know that failing to allow their body to make the appropriate motor adjustments could lead to serious accidents. Even seemingly very simple habitual actions, such as driving a nail with a hammer, closing a door, or raising a glass, call for a series of hardly noticeable adaptations to the characteristics of the object and to a concrete setting.

Whether we adjust the sail of the boat to rough wind conditions or hit a ball with a golf club, immediately we feel the success or failure of our motor performance. As we accomplish these acts, we find ourselves in a dynamic relation to values: right or wrong, well-timed or unsuitable, reached or missed. In his brilliant study on human movement, Paul Christian calls such an awareness of the quality of motor performance "consciousness of value in action" (*Wertbewußtsein im Tun*).[3] This value presents itself during the execution of the movement. If the movement is felt as right and successful, this is not due to its conformity to an ideal or correct way of moving, it is merely sensed as the most appropriate one with respect to the given situation. Christian gives the example of ringing a bell with a rope that goes through a hole in the ceiling into the bell-chamber. The ringer's action is successful only if, without comparing or selecting a motor form, his hand exerts the right pressure on the rope and, sensing the interplay of acting and reacting forces, finds the most suitable rhythm for pulling and letting go and, thus, swaying the bell. He does not know how he does all this; he just allows his body to find the right movement and the appropriate moment of release of the rope. As Buytendijk correctly pointed out, such an intuitive knowledge of values plays a vital role in sport, art, and many other activities. "My body discovers how I have to pull the rope of the church bell; it knows (in its own unconscious way) how the balance can be maintained or recovered or what kind of movement and force should occur in the entire muscle system so that it can give the strongest impulse to the javelin or the discus, just as the hand releases it."[4] This kind of discovery does not lead to an explicit and rational knowledge, and the exact form of the movement cannot be defined in advance. The moving arm creates the form through a vague feeling of the positive or negative development of the action. While throwing a ball into the basket, the player adjusts the angle and the force, one to the other, in such a manner that a slight change applied to one of them implies the necessary modification of the other. A technically good throw is not necessarily a successful one and, conversely, a successful throw does not always display an excellent technical skill.[5]

In all these examples, spontaneity is opposed to reflexive willing: we no longer think or will the movement, we make an unreflective usage of the body by letting it discover and take the most suitable form.[6] Such a surrender to the body's ability to detect meanings and values occurs whenever our movements naturally adjust themselves to the characteristics of the objects or the tasks. While running on a slippery trail, or playing on the piano in a hall with bad acoustics, our legs or hands take a specific form; they are, as Viktor E. F. von Gebsattel noticed, "forming themselves."[7] Our legs move more cautiously or our hands hit the keys more energetically. As we shift from one kind of activity to another, the change is even more notable. When working in the garden, swimming in the lake, or typing a letter, our movements adapt themselves to the task at hand. Each time, while executing a different kind of movement, the body has to modify itself. Gebsattel was not wrong to speak of the working body, playing body, dancing body, fighting body, or loving body. Sporting activities, which consist of the succession of offensive and defensive plays, require just such a shift from one form of bodily conduct to the next one. There is a defensively and offensively playing body. The change usually occurs spontaneously, without special attention or voluntary decision. Thanks to their consciousness of value in action, the players instantaneously modify their relation to the opponent or the ball.

In this context, Gebsattel referred to a distinction, introduced by Straus, between the expressive movement of dancing and the purposive movement of walking. The former develops when the confrontation between subject and object vanishes in space; the latter unfolds in a space defined by a system of directions. Dance calls for participation and merging, whereas the purposive sort of walking consists merely of reaching a goal. Here again, we experience two different ways of spontaneously forming the body. In dance, the motor activity of the trunk is enhanced; in walking, the trunk keeps its stiff vertical position. The increased mobility of the trunk engages the body's organic energies. "The crescendo of motor activity in the trunk accentuates the functions expressive of our vital being at the expense of those which serve knowledge and practical action."[8] Consequently, the enjoyment of the dancer springs not only from the expression of an idea or a desire, and the disappearance of the usual tensions of everyday life, but also from the awareness of, and the trustful surrender to, the bodily feeling for the appropriate form.

The capacity for inventing | Sport activities strikingly illustrate this spontaneous variability, as well as the unpredictable achievements of the body. Martin Seel, in his excellent article on the aesthetics of sport, argues that the attraction and fascination a sporting event exerts on spectators consist in the perception of a successful outcome.[9]

Besides the optimal physiological condition, a remarkable performance requires various technical and tactical skills acquired over a long period of training. Yet, however thorough the preparation is, a spectacular goal or a record-breaking run remains an unpredictable event and only seldom is it the result of reflexive control and careful calculation. Great performances are achieved when the athletes, by adopting an attitude of unconcerned surrender, allow their bodily impulses and powers to organize the movements.

"During an instant or a certain time," notes Seel, "the trained physical body change into an autonomously functioning living body. The intentional action of the athlete change into the unintentional swing of his living body."[10] In fact, it is the enigmatic autonomy of the available body that induces, in both spectators and athletes, an aesthetic satisfaction. Although all the rules and obligations remain valid and are respected, the body seems to step beyond the limits and orientation imposed by the previous training and displays an unexpected virtuosity. Thus, athletes accomplish what lies beyond the level suggested by their training and experience, what, paradoxically, they are "unable to do." And, quite often, they view their own performance with astonishment. Hence sporting activities make manifest an important and highly attractive element, which is eloquently described as *Zelebration des menschlichen Unvermögen* – "celebration of human inability" – an inability of completely subordinating the body to the will and determining all aspects of the motor conduct.

Notwithstanding the intrinsic indeterminacy of sport performances, many consider that the athlete's proficiency consists in the ability to execute a series of complex, predictable and standardized movements. In truth, an auspicious achievement requires more than a conscious effort.

The voluntary decision invariably meets a limit, beyond which the body is no longer a docile instrument. If, as many rightly contend, doping products must be banned from the universe of sport, it is precisely by virtue of their power to push further this limit and, consequently, to silence the natural spontaneity of the body – a body that may offer either satisfaction or deception. Sport remains a meaningful human activity only if it leaves enough room for the manifestation of the imponderable energies of the body.

In any case, the desire to achieve a complete control is illusory. For the countless fans of the *Tour de France*, the beauty of professional cycling is not so much in the technical and tactical expertise but in the unpredictable achievement of the riders. However thorough and "scientific" the preparation for a race may be, dramatic reversals always occur and continue attracting crowds of spectators to the side of the roads. They are enthralled by the unexpectedness and perhaps also by the fact that, at any moment, chaos (the crash of dozens of star contenders) could emerge in an orderly world.

"The living body (*Leib*) completes what the physical body (*Körper*) is able to do."[11] It does it not only through the adaptation and variation of the movements, but also by way of some well-timed innovations. We watch athletes respond to unforeseen challenges with speed and accuracy. To be sure, overcoming a sudden obstacle necessitates the rapid mobilization of the knowledge acquired through training. However, as Ricoeur perceptively argued in his analysis of the spontaneity of habit, the response is not merely the repetition of a learned and practised behaviour, but the launching of a new movement. "In their infinite variations, our flexible habits summon up a frequently disconcerting spirit of appropriateness: a reflection on mental or physical dexterity, on conversation or improvised eloquence, on the skill of living or culture, will show that each time that we parry a new situation we discover in ourselves astonishing resources to which it is wisest to trust ourselves."[12] If the body breaks with habitual gestures, it is due to the "inventiveness" and "capacity for probing" (*puissance d'essai*) concealed within the already acquired skills.

Beyond their ability to deal with disturbances and sudden challenges, athletes are also able to experiment with unusual and original solutions during the decisive phase of their performance. Their body, as Buytendijk has observed, is invested with a subtle sense of what can and should be tried or risked, with a *finesse d'esprit*, with an "inexhaustible creative power."[13] Such a creative activity may embody the connection of two previously unrelated sets of movement. Arthur Koestler proposed for this act of synthesis the term *bisociation*. While association means organizing elements according to a customary set of rules and skills, bisociation is the act of combining previously unrelated dimensions of the experience. The bisociative movement creation arises suddenly and unexpectedly, it is an upward surge from some fertile layer of our body. When, in the closing moments of a ball game, the demand for a rigid pattern of movement linkage and organization decreases, and the conventionality of routine fades away, players are prone to propose a daring and odd solution.

While inventing new movements, athletes consider the motor situation as an action field whose characteristics they evaluate and immediately understand in relation to their bodily capabilities. As beginners, we have an altogether different perception of the moguls on the ski slope than we do after gaining some experience. We see them first as obstacles and only later as helpful means, since our movement, just as much as our sensing, is "a process of coming to terms with the world."[14] We confront the golf course or water currents according to the abilities we have and we make use of them in relation to the challenges and possibilities offered by these natural settings. "When I slip and fall," writes Straus, "I experience smoothness differently from when I struggle vainly to get up from a glazed surface, and still differently when on an icy street I cautiously explore the condition of the surface before putting my foot forward."[15] Qualities and tasks are apprehended in relation to the implicit awareness of the capabilities of our body.

The *finesse d'esprit*, brought to our attention by Buytendijk, is the body's capability to consider a possible complex motor accomplishment (jumping, rotating, or throwing) – a capability consisting of a feeling of correctness and aptness, an intuitive knowledge of what should and can be done in a particular situation. The body seems to know the characteristics of the action field and of its own available and transposable resources. "Each spontaneous behaviour in daily life is regulated according to the situation, but is, at the same time, 'exploratively' *improvised*. The physiogenesis is not determined (like a morphogenesis), but is, to a certain extent, *open* with regard to the range (*Spielraum*) of the possible. This is true in the case of climbing a mountain, driving a car, but also in that of the so-called habitual actions."[16] But to invent and risk, we must not only rely on this fine feeling but also possess the global form of the movement that we want to accomplish. We cannot represent such a possession. It reveals itself in our virtual and concrete doing. Just as musicians know the whole melody and this knowledge allows them to play the piece, likewise the implicit possession of a global form, an "amorphous source" generates the articulated execution of movements.

Eugène Minkowski has noted, in his phenomenological analysis, that the characteristics of spontaneity become better known when we contrast it with other types of human behaviour.[17] Spontaneous actions are neither calculated and artificial nor strained and mannered. Their meaning and value do not depend on functional or utilitarian considerations; they are meaningful in themselves, merely in their direct, dynamic, and prompt manifestation. Not only are playful movements or unexpected combination of forms and ideas spontaneous, but also sudden discoveries, inventions, and exceptional and heroic acts. In just such an immediate and instantaneous manner, long awaited solutions may be found to theoretical and practical problems. Minkowski points out that the sudden emergence of ideas and surprising flashes are the fundamental traits of both spontaneity and a creative life. Indeed, in moments of inspiration, "there is a veritable bursting forth which comes, like lightning, to project its intense and exceptional luminosity on our inner life, without our being able to say whence it comes, and even without that question occurring to us, on the level of immediate data."[18] This bursting forth and emergence originate in the "dynamism of life" – a dynamism that carries us along and asserts itself as a fundamental trait of our body.

Spontaneous actions are deeply personal and authentic, in the sense that they express subjective sensations, impulses, intuitions, and inventions. Because they are prompted only in part by external influences, they can be considered as the true manifestation of our "inward life." The liberation from constraint may happen unexpectedly. First we might carefully follow a set of rules and, suddenly, without warning, we go beyond what we have planned and expected.

Buytendijk reminds us that the principles of efficiency, utility, and selectivity yield, here, to other values, such as exuberance, lavishness, and superabundance.[19] Full of unexpected ornaments and superfluities, spontaneous actions lack functional sobriety and calculated restraint. They display, to a remarkable extent, the deep-seated human desire for constant renewal and qualitative change, freshness, and excrescence.

Minkowski evokes the close relationship between the autonomy of the body and the tendency to perfect the movements in absence of the need to adapt them to a specific goal.[20] Consequently, we are able to concern ourselves with their suppleness, sobriety, or grace. Indeed, graceful movements are perhaps the most successful products of the bodily impetus towards novelty and superabundance. Grace does not have much to do with mechanical precision or technical purposefulness. Rather, it is the result of individual variations and nuances, the entertainment of luxury, the original and qualitative envisagement of different motor options. Grace springs from the ability to express an idea or a feeling through a rich and astonishing spectrum of movement compositions.[21] It requires an intuitive and free improvisation, albeit within the realm of certain intentional structures and relations. In this state, the body is allowed to propose exploratory and useless motions which, in our everyday life and for obvious and necessary reasons, have been eliminated for the sake of uniformity and efficiency. With the introduction of the wealth of different alternative modes of action, the body seeks to exhibit its own internal and unconscious resources and, by making use of them, produce an abundance of forms.[22]

IMPROVISATION | Some of the above mentioned qualities are also constitutive elements of improvisation. What is needed for ingenious improvisations in music, dance, or play is a series of spontaneous actions, consisting of the variation and invention of forms. We tend to contrast improvisation with a calculated and carefully planned action. But the release from conventional and premeditated ways of performing does not mean that the improvisatory behaviour is altogether devoid of order and consistency. In every kind of improvisation we find order and repetition, assuring the preservation of the coherence and unity of the performance.

Children introduce surprising solutions into their play and, at the same time, conform to some self-imposed rules. Musicians propose unexpected melodic solutions while adhering to certain melodic and rhythmic formulas; they are working on models. Bruno Nettl defines the model as a "series of obligatory musical events which must be observed, either absolutely or with some sort of frequency."[23] These events or points of reference are tones, motifs, rhythmic figures, or chords, reoccurring more or less regularly in a given piece of music.

From the concept of model it follows that the improvising performers retain and store certain motifs and return to them at more or less regular intervals. No less important is their capacity to anticipate both new "tonal images" and their consequences. A successful improvisation comprises both the desire to achieve coherence through reaffirmation of already used elements and the readiness or impulse to invent and explore something new.

As Jeff Pressing called to our attention, it is an interaction with a "referent" or a "guiding image" that provides the musician with an impetus for improvisation. "The referent may be a musical theme, a motive, a mood, a picture, an emotion, a structure in space or time, a guiding visual image, a physical process, a story, an attribute, a movement quality, a poem, a social situation, an animal – virtually any coherent image which allows the improviser a sense of engagement and continuity."[24] Musical improvisation evolves, essentially, through a pathic-encounter: the referent seizes the musicians and "invites" them to generate new sound patterns. They explore and invent on the basis of the "continuous aural and proprioceptive feedback."[25]

All good performers display an urge to avoid mechanical repetition and, by keeping in mind the essence of the work, consciously embellish their play with expressive modifications. Some go even further and spontaneously react to numerous momentary variables, making their performance "an exciting adventure." They translate their reaction into the music by introducing subtle "distortions": they play certain notes slightly faster or slower, louder or softer.[26] The most important source of these deviations is perhaps the tone itself. Like the capricious movements of the ball, the playful unfolding of tones holds the performers in their spell, exerts on them an attraction and generates the tendency to improvise. "Their own play is a play with us and our play with them."[27] Why are musicians so inclined to play with tones? Because the resonating tones themselves communicate an "impulse value," a compelling affective appeal to their bodies. It is the pathic character of their momentary auditory experience that elicits a particular response and provides an incitement for the subtle improvisation. Heinz Heckhausen affirms that "music-making may, in a more limited sense, become play when it is the object of improvisation, free variation, imagination. The player himself (!) creates the possible appeal for novelty, complexity, and surprising effect that retroacts on him and facilitates the continuation of formative activity."[28]

That is not all, however. Improvisation also calls for a sense of being carried along, giving and abandoning oneself to both the momentary feelings and the creative impulses of the body. Barry Green claims that during a fulfilling improvisatory experience the guiding model gradually fades away and becomes secondary. At this time, the music seems to emanate from the improviser's body.

Liz began to experiment a little more: she played faster and slower; higher and lower; happily, romantically, and sadly; and observed the ways in which the dancers responded to every change in her mood. Then she slipped out of Bach's style completely and began to play entirely from her own feelings, improvising with different rhythms and harmonic structures. She explained later that she had 'played without thinking' and let the music flow from her moment-to-moment sense of discovery.[29]

The expression "sense of discovery" refers to the improviser's relaxed receptiveness that allows her hands to explore a new range of sounds and unusual harmonics as well as rhythmical effects. It also implies that she accepts the possible discrepancy between a specific intention and a solution arising from the body. The musical forms are no longer produced in the strict sense but discovered, and the resulting music may elicit an effective surprise in the performer.

Nearly all studies on improvisation emphasize the involuntary aspect of the performance. Jeff Pressing maintains that improvisation provokes in the performer "an uncanny feeling of being spectator of one's own action."[30] Alfred Pike believes that the jazz soloist methodically pursues a definite musical goal. Nevertheless, "during the fervour of improvisation, jazz ideas leap to his mind and fingers."[31] It would be, of course, erroneous to view the whole process of musical improvisation as an unconscious, automatic process, freed of all concentration and attention. "The new is rooted in the old."[32] The musician must build up melodic formulae during a long period of learning. In fact, as we have seen, improvisation usually begins and ends with the implementation of previously learned musical ideas. However, improvisatory performance requires more than just a capacity for perceiving, remembering, and reproducing patterns. These are completed with spontaneous inventions, thus making the shaping of melody and rhythm no longer consciously monitored. Now the music sprouts, as it were, from the depth of the body. Without purposeful pre-assessment and planning, the improviser surrenders to a "creative compulsion" that is present in the body.[33]

In his thought-provoking book, David Sudnow proposes not only to explain his progress through jazz improvisation but also to throw some light on the "nature of the human body and its creations."[34] At first, while learning various chords, he found the melodically appropriate scale for a particular chord and came up with a rather hurried combination of these musical formulae. Later, he attempted to fit together chords and melodic fragments. Finally, he reached a stage where he could stop being concerned with a suitable melodic pattern, and play, in a more relaxed and unhurried way, the notes that were not necessarily tied to a specific chord. A melody could be generated and sustained whether it was or was not appropriate for a chord pattern. Sudnow's play was no longer guided by a "rigid time schedule" and an anticipation of chord-specific formulae. He did not have to look "past the hands' ways to their destinations."[35] Rather, he was able to focus confidently on the moving of the hands and let his fingers choose the notes. Any note could have been used for his jazz improvisation. "Good notes were everywhere at hand, right beneath the fingers." A significant change occurred in the improvisatory process when the "melodic hand" was able to dispense with an abstract musical scheme or thought and, by "tasting possibilities," to produce melodies of its own accord. For Sudnow, such a change was accompanied by an experience of automaticity, an absence of conscious controlling and monitoring, a feeling of being guided by the creative capacities of his hand. Thus the hand responded to the produced sound or a momentary feeling, and not to the representation of an abstract musical idea.

One might consider with suspicion, or reject as "metaphysical," the desire to emphasize the vital role of bodily capabilities in generating improvised music, and view both the refinement of improvisatory skill and production of novelty merely as the proper use of the capacity for implementing internal motor programs. Simply put, the improviser operates with various internal models and not with the unreflective movements of the hands.

Eric F. Clarke, who subjects Sudnow's account to some criticism, defends this view. Improvising performers, argues Clarke, cannot "achieve fluency and accuracy" in the absence of abstract representation of "hierarchical structures of motor programmes."[36] They must construct and adhere to a more or less carefully elaborated harmonic outline or pattern. In the absence of the representation of complex "generative structures," the control and shaping of their performance become hesitant and slow.

For Clarke, the motor organization of music-playing is very similar to that of typing. He refers to an experiment designed for typing a text with and without advance information. Those who had no text preview performed much slower than those who had received information in advance. Therefore, "a literal interpretation of Sudnow's claim is untenable," since every improvisatory process is guided by a representation of motor programs.

There is, however, a significant difference between these two types of motor performance. The typist's hands must adapt themselves to a previously fixed series of events. The improviser, on the other hand, does not necessarily have to subject the movements to a preordained framework. Some artists may implement preordained motives; others, however, make no attempt to rely only on past strategies. Whether motives do or do not crop up in the unfolding musical development, the improvisatory process remains essentially open: new notes can be introduced unpredictably and suddenly at any moment.[37] Obviously, such openness does not cancel out the aptitude for coherent and consistent performance.

What Sudnow achieved was precisely a sort of releasing from a rigid commitment to a set of formulae. The notes he played were not imagined in advance, but found effortlessly "right beneath the fingers" on the keyboard. When no text preview was provided to the copy-typist, the motor performance could easily have become inaccurate. In Sudnow's jazz improvisation, however, "wrong notes" simply did not exist; almost any note could generate a good melody. Hence the confident and relaxed attitude: he knew that the solution chosen by the hands could always be used for a good purpose. Thus, the important thing in improvisation is not to develop structures of motor programs and then find a suitable solution, but, as Pressing says, to "accept the first solution that comes to hand."[38] In short, improvisation is not a matter of always hitting the right key or always placing the accent on the appropriate beat.

Clarke's objection presses us to raise an important question: should we consider the representation of movement – the mental image that is obtained on the basis of sensory impressions and previous motor experiences – as an indispensable condition of complex creative performances?

We are surely unable to form a mental image of certain familiar movements such as tying the knot of a neck tie. Therefore, Arnulf Rüssel contends that the basis of several complex motor performances is not a representation but a *readiness* towards the execution of movement (*Bewegungseinstellung*).[39] This expression denotes the inclination to consider the situation and perform the appropriate movements. Creative improvisation may proceed from this readiness and any attempt to focus on how the movement is to be performed, or how the various motor segments are related to each other, could make the performance hesitant and slow.

Paul Ricoeur also alleges that spontaneous improvisation depends on our bodily capacity to probe in all directions and not on the representation of a motor structure. It calls for a surrender to the body's singular potential for invention, variation, and adventure. "In relation to this inventiveness of habit itself, our perceived or represented models play only a secondary or critical role. The guiding scheme creates nothing, only judges the improvisation which approaches what was desired. In this sense it is always true that we do nothing voluntarily unless we have first realized it involuntarily."[40] We all know from experience that, after a certain number of trials, our body suddenly discovers the right or satisfactory way of moving and we do not really know how we succeeded. Such a discovery does not exclude the possibility of looking back and reflecting on the segments of our performance. We may also interpose a representation between our readiness to move and the execution of effective movement. However, such a representation could easily hinder, rather than promote, a creative accomplishment.[41]

spontaneous morality | Why, after all, should we exalt bodily spontaneity? Amid the continuous demands of external conformity and the growing erosion of self-reliance, we have somehow lost the capacity to listen to our feelings, convictions, or desires and to act in accordance with them. We prefer to rely on stereotyped formulations or protective conventions rather than to follow our inner voice. In the final analysis, such a proclivity to conform encourages the gradual erosion of a key aspect of our moral fibre – the intuitive feeling that distinguishes right from wrong. No doubt our feelings need education; the more we live by them the more they offer us appropriate guidance.

When we yield to our feelings and rely on our spontaneous know-how, we are truly in contact with ourselves. This contact can give rise to what we might consider as unpremeditated good actions. Without a moment of hesitation, without considering our own interest, we might bring help to a person in distress. Or without any lengthy moral reflections, someone stops doing the task at hand and walks with us to show the direction we seek. "Actions like that," writes Robert Spaemann, "make life worthwhile."[42] We can see in such an action, emerging from internal sources, the foundation of altruism. Through spontaneous movement, our fellow human beings often bring out what is the best in them.

"Ethics can never be purely and simply an affair of the mind," correctly states van Peursen.[43] It is not merely a matter of rationally understanding abstract arguments or values. It is an exercise in practical reasoning. It displays a qualitative distinction grounded in our feelings, and calls for the readiness to act in accordance with an affective appreciation of values. "Spontaneity is situational," says Eliot Deutsch; "it is a sensitivity to what is demanded by, and is at stake in, a particular context; it is not a reliance upon an inner sense of what is right/wrong in the abstract or upon an internalized, superegolike assimilation of an imposed rule."[44] We act not by relying on fixed formulae, but in accordance with our affective participation in the meaning of the given situation. Because of this, the notion of authenticity, the contact with our feelings, has great importance in any environment that encourages external dependence, fixed habits, and conventional living. Conversely, through spontaneous actions, we are better able to live according to the ideal of authenticity and, thus, gradually strengthen our moral sense.[45]

It is also worth noting that spontaneous movements help to foster another aspect of our moral life: the fine sense for nuance, delicacy, timing, and proportion – in brief, tact. Important social virtues, such as politeness, thoughtfulness, civility, or benignity, are rooted in this ability to perceive and value what is unique in a situation at hand and outside the realm of the utilitarian and impersonal.[46] A mature practice of morality depends upon the capability to develop sensitivity towards the prompt and gratuitous stylization of forms of personal interaction, however superficial they seem to be. Morality is not simply a matter of obedience to some externally imposed rules; it calls, Helmuth Plessner tells us, for a behaviour inspired by enthusiasm and a "spirit of luxury."[47] The heart of ethics lies above and beyond all rigid regulations. It is grounded in our luminous and creative intuitions, in our keen sense of the exceptional and, accordingly, relates to spontaneity in a broader sense.[48]

4 | IMITATION

THE MIMETIC BODY | We make use of our body's mimetic aptitude at a very early age and continue to rely on it throughout our whole life: we acquire skills and habits, learn a great variety of movements and re-enact, consciously or unconsciously, various gestures, in the most diverse situations. We all have found great pleasure in producing a caricature of our parents or teachers or employers. Prompted by the mixed feelings of affection and defiance, we enjoyed exaggerating their stiff habits and having the impression of exercising some control over these authoritarian figures. We were eager to indulge a deep-seated desire to make our body look like another's. We liked to reproduce the accent, manners, and gestures of those we admired. How often we yearned to look older and act like adults by simply lowering the pitch of our voice or changing our facial features. We also liked hiding behind masks that we had carefully created.

Helmuth Plessner related "the delight of self-concealment, make-believe and disguise" to the fundamental human urge to imitate.[1] Play is manifestly one of the human activities that offer the earliest occasion to satisfy this urge. Our playful activity does not necessarily have to be tied to a purpose or a motive; we play for the sheer joy we find in impersonating something or someone.[2] Buytendijk observes that, at a certain age, children tend to develop a "mimetic culture" and even approach their surroundings with a "mimetic mania."[3] Both the accomplishment of movement itself and the possibility of becoming, momentarily, a train, an aeroplane, a soldier, or a doctor give rise to a feeling of satisfaction. While imitating, children adapt their own postures and movements to the salient features of the motor situation. They also change roles with ease and rapidity and, by doing this, they are spontaneously forming their body. The reproduction of a figure is done not only with their hands and arms, but also with their legs, torso, head, and voice; their whole body becomes an "organ of mimetic expression" (Horkheimer).[4]

While in front of an audience, and knowing that a good performance on stage always requires an embodied presence, actors skilfully move their bodies to impersonate various characters. They can hardly achieve success without the able movements of their face and hands, the constant change of their body's positions and tensions, or of their manners of walking and speaking.[5]

The mimetic element is also central to dance. In his studies on the anthropological function of imitation, Walter Benjamin has pointed out that, together with speech, dance is rooted in the basic human ability to perceive and reproduce similarities. The "oldest function" of dance was to recognize and imitate animals, natural events, or mythical stories. Drawing on this unique potential, the ancients were able to establish cognitive and expressive correspondences: by adapting their own behaviour to some external realities, they were able to gain a sympathetic understanding of the observable and non-observable world. Consequently, in former times, children's "mimetic genius" – their "life-determining" ability – had to be awakened and consciously nurtured.[6]

Acting, dancing, and playing are all-important parts of rituals. Through ritual practices, children express their individuality and, at the same time, gradually incorporate the modes of moving and being of a particular community. Both the individual and social aspects of an identity develop through ritual practices. The imitation of the movements executed by the older members of the community shapes the children's body: modes of speaking and expressing feelings mould the muscles of their face. Ceremonial walking, sitting, and kneeling also leave a lasting mark. If the previous generation transmits particular values and experiences to younger ones, and their bodies incorporate these, the process of memorization is enacted by mimetic appropriation. It is through ceremonial songs and dances that rituals transmit the embodied memory of the community.[7]

How does the body imitate a person or an object? In most instances, children do not seek to faithfully reproduce a perceived form. They focus on the salient aspects of an activity and repeat them with various rhythms and intensities. If they imitate an aeroplane, they will walk forward, lean their trunk sideways and extend their arms. Ernst Cassirer emphasized that the act of reproduction presupposes an inner production of a model, composed from the constitutive traits of an object. Imitation "never consists in retracing, line for line, a specific content of reality, but in selecting a pregnant motif in that content and so producing a characteristic 'outline' of its form."[8] Its central feature is not the repetition of "sensuous characteristics," but the formative grasping of an object "in its structural relations."[9]

It would be erroneous, however, to ignore the occasional interest in minute details. In some play, children prefer to represent the object's specific peculiarities rather than its quintessential traits. The reproduction of details presupposes the ability to analyze and reconstitute a given reality, and a fine sense of motor precision and control.

Margaret Mead observed the behaviour of the Balinese watching with interest the play of two children or a cockfight.[10] The spectator's two hands, executing two independent forms of mimetic movement, become "separate symbols" of the two moving beings. The re-enactment is so accurate and thorough that the sheer observation of the hands' "dramatic counterpoint" makes one able to follow the fight.

Watching creative artists at work, Mead tells us, we are able to witness such a contrapuntal behaviour. "So when a painter was working with one hand and the other lay on the table unused, it was sometimes found that that second hand provided the more interesting series of postures, as if the neglected hand were playing out a little counterpoint of its own."[11] Similarly, the conductor of an orchestra may indicate a specific rhythmic pattern with the right hand and mime the music's emotional content with the left. Orchestra members are disconcerted when the conductor unintentionally communicates conflicting or ambiguous cues with his technical and expressive gestures.

CONVERSATION | Because of their remarkable flexibility and mobility, the hands exhibit a natural propensity for mimetic behaviour. During a lively conversation, the hands' movements naturally supplement the production and flow of the spoken words. The gestures have sometimes no other purpose than to fill the periods of silence that inevitably occur; they are, as David Abercrombie aptly called them, "silence fillers."[12] If we are inspired or seized by a strong emotion, we tend to introduce, between two words or at the end of a sentence, various unpremeditated "dramatic" gestures. These may serve to elucidate or give more emphasis to the meaning of words, or to reinforce the content of our communication effectively. Ludwig Wittgenstein did not fail to notice this remarkable correlation between our bodily movements and verbal expressions: "How curious: we should like to explain our understanding of a gesture by means of a translation into words, and the understanding of words by translating them into a gesture. (Thus we are tossed to and fro when we try to find out where understanding properly resides.)"[13] In some verbal exchanges, these demonstrative movements prove to be indispensable; it is only through the observation of the partner's hand or face that the meaning of sentences can be correctly understood.

A genuine spontaneous conversation, as Abercrombie says, "is sometimes unintelligible, and it is always illogical, disorganized, repetitious, and ungrammatical."[14] If we were able to read the written transcription of our own conversations, we would find them "horrifying" and "illiterate." Compared to the written prose, conversational sentences lack completeness and fluency. It is no wonder that, consciously or unconsciously, we tend to correct these shortcomings and, through phonetic and bodily activities, spare no effort in making our speech more intelligible; we introduce variations in our voice intonation and tempo, we repeat words and pause between sentences, we make an implicit reference to a material context. We also come to the rescue of our non-fluent spoken language with demonstrative movements; the hand's elaborate mimetic gesticulations contribute to the clarification of linguistic meanings.[15] Thus, while describing to someone a recent event, we accomplish imitative movements, and these are prompted more by the words we utter or the images we conserve than by some external realities we see or hear. Our mimetic gestures depict meanings and ideas in their sensuous concreteness.

Clearly, an informal conversation involves much more than just the exchange of information. We find ourselves facing another living body: we are sensitive to the distance between ourselves and our partners, we see the stiffness or casualness of their posture, watch the motion of their eyes and mouth and perhaps even smell the odour of their hair or clothes. There are other muscular changes and postural tensions that we hardly remark at all. Nevertheless, we are bodily present in a face-to-face communication, and our various sensory impressions constitute an indispensable complement of what we hear and say.

Beyond the hand's versatile dance, the facial mimetic movements play a conspicuous role in a conversation. As the participants speak, their lips and eyes are constantly moving, their skin and underlying muscles change in tension. The complex and intricate motions of the muscles provide them with the feedback necessary for the proper progress of verbal exchange. José Ortega y Gasset compared the body of each interlocutor to "an unstinting semaphore" that is constantly sending various indications or suggestions.[16] Jeremy Campbell studied such a "conversational waltz," during which the speaker and listener, by sending out bodily signals, guided each other: "By means of nods of the head, gestures of the body, smiles or frowns, the listener keeps the speaker informed on how his message is being received, whether it is time he stopped talking, or whether he should continue."[17] Conversely, it is the relative immobility of the listener's face, the absent look, and not so much the lack of desire to communicate, that stifles a conversation, as it happens in situations governed by fear and suspicion.

In his study of the human face, Harvey B. Sarles also called to our attention this aspect of human communication: "From a dynamic point of view, language is a set of muscular movements, whose complexity and beauty match that of the fine dancers and athletes. Like a study of the hand movements of a skilled dentist or musician, the more one studies these processes, the more one is impressed by the capabilities of the human form."[18] The face is a dialectical mimetic surface: it can translate both the content of, and the response to, a message. Similarly to the hand, facial features – grimaces, frowns, mouth motions, glances, and breathing frequencies – can offer concrete demonstrations of feelings, images, and ideas; they have an "illustrative value."[19] For example, slight changes of facial muscles can tell a great deal about the progress and outcome of a sporting event that we verbally describe, and about the reactions – question, doubt, acquiescence, or astonishment – that our description elicits.

Another remarkable aspect of conversation is the imitation of the interlocutor's facial expressions and the particular mode of utterance. When two people are intensely involved in a discussion, they tend to reproduce not only the use of certain terms, concepts, and even speech dialects, but also facial habits, tensions, and movements. We are all aware of the considerable diversity of faces. Georg Simmel correctly remarked that a face strikes us as the symbol of an "unmistakable personality."[20] Yet, not only children's faces resemble those of their parents, but husbands and wives, after living together for many years, also tend to look alike. Members of the same linguistic community share a certain number of facial features. The common propensity to develop the same or similar muscular habits shapes a distinct "facial dialect."[21] If these facial surfaces, each expressing a "definite spiritual individuality" (Simmel), are brought into a unity, this has to do not only with the form of the skull, but also with the similar ways of using speech muscles. Remarkably malleable, the face is a bodily part on which emotions and memories, as well as specific verbal activities, leave lasting traces.

Verbal communication, consisting of the production and imitation of an almost infinite variety of sounds, is deeply rooted in some bodily functions. The linguist Iván Fónagy considers sound expiration as a "specific bodily behaviour," one of the most important motor activities.[22] Our "vocal gesture" reflects the rhythm of respiration, muscular tension, and the position of our body. The laborious process of learning a language already illustrates this correlation. Children grasp and reproduce the prosodic features of a voice – its intonation, intensity, rhythm, and expressive articulation – before understanding the precise meaning of the words.[23] They take pleasure from repeating sounds that have no significance whatsoever, provided that they are "interesting." Their sole aim is to make the captivating activity last.[24] At a later stage of their life, they are no less delighted to reproduce the vocal utterances made by adults. During a conversation, adults also repeat certain words in order to emphasize selected ideas or signal their attentiveness. They might do this only for the pleasure that the flavour of a curious combination of syllables gives them.

While the vocal imitation of children has been repeatedly studied, less attention has been paid to the imitative sounds produced by adults when they communicate with a child. Melodic speech is one of the first discoveries of children. The flow of words they utter is more melodic than that of adults. It consists of a higher degree of intrasyllabic tonal movement. When adults speak to children, they tend, through an effort of identification, to spontaneously mimic the melodicity of the child's speech.[25] Such a "regression" in the way of speaking and acting is very beneficial for the children's growth since it creates a bridge, through which they are also able to identify themselves with the adults.[26]

The same adaptation may occur during an ordinary conversation. If the participants are willing to empathize with each other and follow, with attention, the conveyance of thoughts and desires, the tonal movement of their voice naturally adapts itself to the quality of their partner's speech. Because of the reciprocal sharing of the other's emotional attitude and communicative intentions, the produced sounds switch from a low to high degree of melodicity and vice versa. Conversely, the purposeful imitation of speech patterns could lead to a more empathetic approach and communication. A variation in speech melodicity is less significant when the speakers relate to each other with some tension. In such cases, their voice intonation becomes even, as happens, for example, when a verbal exchange turns into a scolding or an order, or the prevalent aggressive atmosphere suppresses the possibility of mutual understanding.

the involuntary in imitation | Imitative movements may either accompany or follow the perceived motor event. The term *echokinesis* refers to an imitative behaviour that follows, with some delay, the perception of sound patterns or visual forms.[27] There is a deliberate intention to produce similarities. Both actors and children perform echokinetic types of imitation when they copy previously perceived or represented movement structures. Although, as mentioned above, the movement is carried out with identical means, they make no attempt to faithfully reproduce a given action.

Synkinetic movements arise in synchrony with perceived or imagined events such as the rhythmic movement of a dancer, the melodic pattern of music, or the sudden and dangerous acceleration of a vehicle. A common feature of these examples is the suggestive power of an object: it elicits concomitant movements. Many of us have assisted at an exciting sporting event and found ourselves unable to sit still. We recall how easily our hands or legs followed and anticipated movements; we felt as if we were compelled to help the moving player to reach the intended objective.

An important element of synkinetic imitations is the interest with which spectators follow a perceived event. If the sensitive fingers of the Balinese watching a fight move nervously and independently, it is due to the characteristics of the motor experience itself; the fight, with its dramatic twists, exerts on the spectator a fascination, and provokes his active participation. As long as it remains a captivating experience, the Balinese can hardly prevent his hands from trembling, twisting, and twitching with animation.

Synkinetic movements occur regularly when we approach a conversation with sincere curiosity and intensity. Beyond the voice, which adapts itself to the pitch and melody of sounds, the different parts of our body execute concomitant movements: we bow our head, lift up our arms, and move forward or backward in synchrony with our partner. As we accomplish hardly perceptible small gestures, we indicate to our partners, most often without thinking, that we are, if not in complete agreement, at least in tune with them.

How strong is the involuntary need to reproduce a perceived movement?[28] Doubtless, in some situations, the powerful compulsion to imitate is so strong that the synchronized movement of crossing our arms or yawning occurs almost automatically. However, momentary feelings, disposition states, or preoccupations may minimize this primary motor influence and hinder or delay imitative reactions. As Kurt Kofka pointed out, we are less inclined to surrender to the immediate experience and respond to someone's laughter when we are tired and overtaken by a gloomy mood.[29] He argued, therefore, that the incapacity to resist a suggestive power is less conducive to imitation than is the conscious and controlled readiness to make ourselves similar to someone or something.[30]

Notwithstanding the insistence on the primacy of voluntary control, the vivid and stirring quality of dynamic appearances does affect us on many occasions and prompts a great number of mimetic responses. We seem to spontaneously imitate a rhythmic movement if we perceive it to be visually attractive. While dancing, we are keen to surrender to the invitation of a visual or auditory perception and translate it into movements. Motor responses, released under voluntary control, do not necessarily diminish the observable fact that we tend to satisfy the powerful need to recreate what we perceive. Once, during a job interview, I found myself reproducing, involuntarily and without any delay, the movements made by the director of personnel: I was compelled to mime his crossed legs and interlaced fingers, and thus expressed my interest in the position and eventual willingness to co-operate.

The reason for this is that we perceive both the uttered sounds and the speaker's bodily gestures "in muscular terms."[31] The form and quality of an appearance elicits in us a strong impulse to move. We all know from experience that when we perceive a ball moving towards us, we feel a barely resistible urge to catch it or kick it back, as well as pleasure in doing so.[32] Similar experiences occur in museums: the sign "please don't touch," or the restraining ropes, would not be installed if people were able to easily suppress their strong urge to make contact with the objects on display. Antique furniture, machinery, and weapons arouse in the visitor the strong desire to follow their shape with gentle caresses. If our smiling colleague moves towards us with an extended right hand, we respond without any hesitation with the same motion. How can we do otherwise? We find it so much easier to yield to the impulse to shake the hand and reciprocate the smile than to keep our hand immobile and stand motionless with a stern face.

Herbert Read and Rudolf Arnheim have advocated that to appreciate sculptures, both visual and tactile contacts are required. Sculptures are not only made by the touch but also for the touch. Unlike the detached and distant images hanging on walls, sculptures belong to our own life space and, as such, create "a spontaneous intimacy" with our body.[33] Once again, we not merely enjoy, but are induced, to respond to the invitation of their massive presence, especially if they are placed on our own level. We feel a curious desire to follow, through the gentle movements of our hand, the contours of the three-dimensional human figures.[34] We might feel a similar urge to accomplish movements while glancing at a distinctive architectural structure. "Looking at the Eiffel Tower," writes Jacques Lecoq, "each of us can sense a dynamic emotion and put this emotion into movement. It will be a dynamic combining rootedness with an upward surge, having nothing to do with the temptation to give a picture of the monument (a figurative mime). It's more than a translation: it's an emotion.... In fact we constantly mime the world around us without realising it."[35]

All these miming movements unfold in the presence of objects and events that somehow speak to us; their properties emit subtle messages and "invitations." Following the insights of Kurt Lewin and James J. Gibson, Rudolf Arnheim speaks of the "demand quality" or "affordance" of both works of art and the objects we use in our everyday life. We experience their shape, size, and colours as attractive or repulsive, inviting or forbidding. We become attentive to their expressive qualities, a life that stirs our feelings.[36] As long as we approach them with due respect and interest and remain receptive to their "motivating quality," we will feel the urge to respond to them with bodily movements. If we see a very old book, we find it almost impossible not to open it and gently put our hand on one of its pages; we know that the perception of the object remains incomplete as long as we do not establish a tactile contact with the faded papers and recreate with our fingers its configuration.

Children, sensitive to the physiognomic qualities of forms, are prone to spontaneously repeat words or imitate the movement of animals (snake, bird, or ape). They are attracted to the motor-affective meaning that they discover in a living being. Jean Piaget believes that this interest in a form depends on the sheer possibility of reproduction. All living forms become significant for a child as long as they are susceptible to repetition. It is the already existing "sensory-motor schema" (an inner structure constructed out of perception and movement), that inspires children and evokes in them the need for imitation.[37] Repeated for its own sake, pleasurable in itself, imitation tends to occur in a relaxed and serene atmosphere. The absence of restriction helps children to focus. Older children, for example, more willingly imitate a person than they do an object. The latter is usually apprehended in a practical and utilitarian context and gives rise to goal-oriented movements. Perceived without practical concern, the former – human figure – tends to elicit imitative gestures.[38]

Awareness of the Body | The non-reflective understanding, the ability to immediately apprehend the meaning of movements and forms, is another important element of imitation. We not only take part in a particular intention and "affordance," but, thanks to our body's comprehending power, recognize it as our own possibility: we are able to produce a similar intention or dynamic force.

The communication or comprehension of gestures comes about through the reciprocity of my intentions and the gestures of others, of my gestures and intentions discernible in the conduct of other people. It is as if the other person's intention inhabited my body and mine his. The gesture which I witness outlines an intentional object. This object is genuinely present and fully comprehended when the powers of my body adjust themselves to it and overlap it. The gesture presents itself to me as a question, bringing certain perceptible bits of the world to my notice, and inviting my concurrence in them.[39]

Merleau-Ponty holds that an expressive gesture finds its "equivalent" or "confirmation" in our own body. The "equivalence" implies that the body itself has a "primitive self-representation" (Meltzoff and Moore), a basic awareness of its own abilities and impulses. Doubtless, such a bodily self-awareness is the starting point for all subsequent self-objectivation, but the body must first make possible the apprehension of the meaning of an expressive reality and its instantaneous imitation. Merleau-Ponty's example of the baby reacting to an emotional expression well illustrates the immediate understanding of others and the rudimentary grasp of one's own possible movements. Without previous experience, a fifteen-month-old baby will open his mouth if we playfully take one of its fingers between our teeth and pretend to bite it. A spontaneous bodily response occurs because the baby feels its own mouth and teeth from inside as "powers" of envisaging similar intentions and producing similar effects. "It perceives its intentions in its body, and my body with its own, and thereby my intentions in its own body."[40] Thus, the baby perceives correlative intentions and, simultaneously, senses its own motor abilities. Both the experience of the expressive gesture and the nonreflective recognition of the ability to accomplish similar movements make possible the imitative behaviour. The imitation of expressive acts requires an internal relation to bodily abilities. In his lectures dealing with the roots of language, Merleau-Ponty stresses this cardinal aspect. Imitation "presupposes that he (the child) experiences his own body as a permanent and global power capable of realizing gestures that are endowed with a certain meaning. This means that imitation presupposes the apprehension of a behavior in other people and, on the side of the self, a noncontemplative, but motor, subject, an 'I can' (Husserl)."[41]

Helmuth Plessner, in his anthropological analysis, puts forward somewhat similar views.[42] He claims that imitation is directed by the reciprocity of bodily schemas, which are the concrete internal representations of the structure and ability of different parts of the body, based upon previous experiences and related to a specific motor context. The exchange of looks allows us to become aware of the reciprocity of both the points of view and the possibilities of behaviour. Dancers, painters, sculptors, and playing children are particularly endowed with the gift of "absorbing movement by looking at it" (Merce Cunningham).[43] Without detailed instructions, they are able to compare and select motor forms, seeing them as their own.[44]

Let us now consider, more closely, the awareness of a possible intention and ability in one's own body. When we represent another person, we relate to an "inward attitude." This essentially consists of a particular way of being, in the sense of being friendly, orderly, artistic, religious, philosophical, and so forth.[45] We recall that Ernst Cassirer speaks of a "model," this being the outcome of a "creative and formative activity," an "inner production"(*Vorbilden*), carried out by the human spirit. Here we may evoke the notion of an internalized role or pattern that suggests to us how to act in a particular situation.[46] Once the attitude, or model, or role is actively formed and strengthened, it becomes the regulating guide for the execution of movements. Without much effort, we imitate the conceited professor's speech or the disciplined soldier's rigid posture. Our verbal utterances and motor behaviour do not necessitate specific representations and commands; we do not have to deliberately evoke a particular way of walking, looking, and shaking someone's hand in order to execute them. Our mimetic gestures spontaneously spring from the already formed inward attitude. "When we adopt an attitude," writes Jürg Zutt, "that corresponds to the idea of an unfamiliar being, we notice in ourselves a psychical fact from which, while our ego can look upon it, our gestures emerge with a perceptible autonomy."[47] Even the seemingly unimportant movements emerge from this autonomous inward character. Thus, actors are able to respond appropriately to any unexpected events that might occur on stage. Their bodily autonomy allows them to execute movements that, nevertheless, bear the mark of a recognizable style – a style shaped by previous experiences and adapted to the characteristics of the actual circumstances. They do not need to select and guide their movements, just as much as they do not have to have a clear and accurate conception of the inward attitude in place. The latter is no longer represented but lived as a constitutive element of their whole being: for the time being, they *are* what they play.

There is a reciprocal relationship between an acquired inward attitude and the corresponding motor behaviour. The former gives rise to imitative movements, the latter, in its turn, reinforces and amplifies a way of being. In acting, the careful practice of movements helps to discover and feel the meaning of basic dispositions.[48] If we keep performing acts of worship, our faith will surely be fortified. Max Scheler has rightly noted that the religious consciousness is not wholly developed independently of bodily expression: "ritual is an essential *vehicle* of its growth."[49] The same holds for other kinds of disposition or casts of mind: the more we carry out some gestures, the more our inward attitude comes to shape our whole being. "The professor putting on an act that pretends to wisdom, comes to feel wise. The preacher finds himself believing what he preaches. The soldier discovers martial stirring in his breast as he puts on his uniform."[50] The outward presence and action bolster one's guiding model and result in the embodiment of a social function. "Normally," remarks Peter L. Berger, "one becomes what one plays at."[51] The process of personal appropriation usually does not follow a deliberate plan; through our involvement in an activity, we inevitably grow into a certain way of life and, conversely, by holding on to some beliefs, values, or goals, we effortlessly behave in a certain way.

The second segment of this correlation brings us to the ideo-motor phenomenon.[52] Ideo-motor actions are movements that are generated and guided by ideas derived from the perception or expectation of a reality. The movement may immediately follow the idea (synkinesis) or occur only after a delay (echokinesis). In both cases, it is an idea that elicits in the body an involuntary response. In this connection, we may refer, once again, to the imitative movements that we execute while listening to a musical piece or watching an exciting sporting event. Since, as we have already observed, a conversation is a bodily event, the ideas that we form following the perception of the sounds and the understanding of their meaning also provide the impetus for movement.

Why does an idea elicit a motor reaction? Why is it a cue to movements? Just like tones easily provoke motor responses, certain ideas exert on us such a powerful influence that we immediately display a motor reaction. We can hardly suppress a movement when we think, for instance, of being hit by an approaching heavy object. The relationship to this thought is just as much a pathic experience as is the encounter with a ball or a strong smell. Because some ideas, as much as an image or a plaything, have an impulse value; they induce both feelings and movements. Alfred North Whitehead was perhaps referring to this kind of induction when he defined thought as a kind of inspiration, which makes us jump up from our desk in rapture: "A thought is a tremendous mode of excitement. Like a stone thrown into a pound it disturbs the whole surface of our being,"[53]

The impulse seems to be particularly strong in good acting. If the director suggests an idea, a subtle inner movement is provoked in the actor's body. The same response occurs in all of us when, for example, we are asked to imagine the loss of a loved one. It is more manifest in acting since, as Peter Brook observed, "the actor is a more sensitive instrument and in him the tremor is detected."[54]

sympathetic communication | We are able to accomplish very satisfying performances, in all areas of our life, if, in a relaxed manner and by forgetting all our worries and fears, we submit ourselves to the compelling influence of an idea or a symbol. William James, who discusses the ideo-motor phenomenon in his writings, argues that excessive concerns neutralize the exciting effect an idea can exert on the body. "Stated technically, the law is this, *that strong feeling about one's self tends to arrest the free association of one's objective ideas and motor processes.*"[55] Therefore, James, preaching the "gospel of relaxation," advises us to silence all "egoistic preoccupation" and surrender ourselves, in a state of absolute unconcern, to the ideas' dynamic influence.[56]

In fact, all imitative movements call for such a "positive approach" (Buytendijk). Performers of modern dance, for example, take a heightened interest not only in the expression of their inner state, but also in the characteristics of a natural or human milieu. There seems to be no hiatus between a dancer's "interiority" and the various aspects of the movement. And the rhythmic succession of movements seems to be guided by the sympathetic perception of a particular space.

Ritual ceremonies also allow us to open ourselves to, and communicate with, something meaningful and greater than ourselves, in both space and time. Although we may focus on past or future events, rituals make us fully aware of the present. Whether we take part in a church ceremony or an athletic contest, we are no longer subject to the constraints usually prevailing while we execute a purposive action. We become fully attentive to the immediate experience and allow ourselves to be moved by it.[57]

In his analysis of Benjamin's theory of mimetic experience, Jürgen Habermas speaks of the "uninterrupted connection of the human organism with the surrounding nature."[58] In a similar vein, Ulrich Schwartz believes that Benjamin's theory comprises a fundamental anthropological assertion: "The mimetic faculty first makes possible the experience of the world in an empathic sense."[59] The various aspects and qualities of the environment are no longer perceived in confrontation, but accepted with a sense of involvement and participation. Thus, the mimetic capacity includes not only the gift of producing similarities, but also the bodily potential on which we draw in order to act in unison with the world and to perceive it with sympathy and care.

The mimetic experience is, after all, a mode of knowledge. It may be considered just as vital for our well-being as the scientific mode of knowing. Objects are perceived in relation to each other and with all their individual, dynamic, and affective qualities. They resonate in us and induce us to recreate their dominant elements. We then sing what we hear and dance what we see. We do not seek to control or extract facts from their context. We approach everything in a relaxed manner, letting them be as they are and entrusting ourselves to their suggestions and demands. Such a deep sense of kinship particularly nurtures an organic outlook that recognizes the interplay of global and dynamic processes in both the human and natural habitats.[60]

5 | Rhythm

INTERACTION RHYTHM | The title of this chapter might suggest an analysis of the body's biological rhythms such as the circadian cycle of sleep and wakefulness. Though the periodic physiological processes are fascinating subjects of study, it is not my intention to investigate how the body's inner clock regulates and guides some of our daily activities. These issues have received increased attention in recent years; students of chronobiology and chronotherapy have published important contributions that lead to our understanding of the body's capacity for time keeping and healing.

I intend to address the following questions: How does the body generate rhythmic movements? What are the main characteristics and significance of rhythmic motor performances?

Similar to imitative actions, children execute rhythmic movements from a very young age. Their moving bodies are, in Nicolas Abraham's words, "rhythmizing from the start."[1] The early rhythmic attunement to the mother's voice and movement provides infants with a highly satisfying emotional experience.[2] Affective communication through sounds and gestures is essential, allowing for the child's learning to take root and steadily progress in the society. Close interaction between vocal expressions, movements, and affectivity continues to play a central role in many subsequent bodily activities such as singing, dancing, and ritual playing. The fundamental desire to adjust to the perceived rhythm will also remain a pivotal feature of many human interactions.

Students of human kinesics – those who explore the role of body movement in human communication – have shown that people interacting with each other display a natural tendency to adopt their partner's rhythm. Either the whole body moves in synchrony or only one of its parts. This "interactive dancing" usually remains unnoticed. If the partners suddenly became aware of the synchronized movements, their communicative behaviour could become disturbed. While, as Edward T. Hall pointed out, the tendency to synchronize the movements is innate, the characteristics of motor rhythm itself are cultural.[3] Not only do people of different cultures move faster or slower but they also stress sound and movement segments in a particular manner. Moving from one culture to another brings with it the challenge of manifold learning, one of them being the adaptation to new, subtle rhythmic patterns.

In our everyday life, we share and respond to a great variety of rhythms. We dispose of a repertoire of rhythmic patterns, from which, in a given situation, we select, often unconsciously, the most appropriate one.[4] The context, as well as the nature of our activity, provides the script for the suitable rhythmicity. We may become conscious of these ongoing adjustments when we see someone unwilling or unable to "shift rhythmic gear" according to the requirement of the context. Entering a church or a museum, we immediately take up a different behaviour pattern: we move with a slower pace and make frequent stops. Workers, however, called in to repair or replace something, do not alter the rhythm of their walk or speech. They surely move in synchrony with each other, but fail to react to the new circumstances. In a foreign land, those who fail to adopt the local rhythm are immediately recognized as tourists or visitors.

Conversation calls for a continuous adaptation to the demands of a rhythm introduced by the partners.[5] The rhythm is established and understood not only through the recurrent sound pulses, but also through a great variety of bodily movements such as the nods, smiles, frowns, and slight touches. The chosen rhythm may serve several purposes. It may help to predict what comes next, hold the partner's attention, display an immediate non-verbal desire or reaction, or strengthen a bond between the speakers. By its nature, a genuine conversation is spontaneous and undetermined, even though it unfolds according to some kind of order. Beyond the selected themes and the partner's willingness to listen and communicate, it is the commonly adopted rhythm that brings coherence to the exchange of words.

Rhythm is certainly a central element of the communication between individuals making music together, illustrated by a sonata recital. As the musicians interpret the part assigned to their instrument, they might play notes slightly faster or slower, or place more or less stress on them. The freedom of introducing subtle rhythmic variations is, of course, not unlimited. Each player has to take into account not only the composer's indication, but also the execution of the co-performer. The interpretation can unfold only if the pianist concurrently foresees what and how the violinist is going to play, and conversely, the violinist anticipates how the music created by the pianist will unfold. It is on the basis of memory of intervals that the players are able to anticipate how the music evolves.

I have already referred to Alfred Schutz's observations on chamber music performance. They make clear that a successful anticipation requires not only the hearing of these intervals between tones but also the seeing of the fellow musician's bodily expressions.

The other's facial expressions, his gestures in handling his instrument, in short all the activities of performing, gear into the outer world and can be grasped by the partner in immediacy. Even if performed without communicative intent, these activities are interpreted by him as indications of what the other is going to do and therefore as suggestions or even commands for his own behavior. Any chamber musician knows how disturbing an arrangement can be that prevents coperformers from seeing each other.[6]

Although not as precise and compelling as the sound, the continuous visual contact contributes significantly to the successful synchronization of an individual interpretation. Seeing the bodily movements allows the musicians to identify with each other, "tune-in" to one another, and generate or follow a particular rhythmic order.

We may compare the musicians' "mutual tuning-in relationship" to the interaction between two dancers: each partner is simultaneously, or successively, leader and follower. Thus the role assumed by one of them is not always obvious. Both may either impose or react to a rhythmic pattern; they move while being moved. This unity of activity and passivity, the expansive and adaptive movements, constitutes the primary characteristics of all play. Whoever plays is, at the same time, being played. The very first encounter – between the infant's lips, tongue, and hands and the mother's body – unfolds according to this very same structure. His subsequent play with a ball is governed by the same complementary dynamics of moving and being moved. Such a structure, comprising the interplay of initiative and adaptation, regulates many other kinds of human activities.[7]

Clearly, we not only send various rhythmic signals, but also adopt the subtle rhythmic suggestions coming from others. We are able to execute synchronized gestures because our rhythmic sensibility consists of identifying ourselves with some temporal sequences and we retrieve them either concretely or in imagination. When we do this, we group together the temporal segments or phases of the movement and emphasize some of their moments. To perceive rhythm we must have the capacity to group recurrent impressions and articulate patterns with an accent. To be effective, this fundamental disposition does not have to be conscious; it functions best if it becomes a motor habit. We perceive a rhythm with our body: we yield to an inner need to respond to this rhythmic pattern with a movement that involves grouping and emphasis. Or we accomplish virtual movements if the perception of rhythm does not provoke in us actual muscular contractions. With or without these contractions, however, the compelling synchronization calls for the anticipation of both impressions and motor responses.[8]

In play or conversation, the rhythmic movements that spring from an inner need concretely illustrate that "the body is essentially an expressive space."[9] Indeed, during these activities, we reveal our distinct traits, our unique way of moving, talking, and gesticulating, in short, our style. "Every human being," writes Kurt Goldstein, "has a rhythm of his own, which manifests itself in the various performances, but of course in various ways, yet in the same performance always in the same way."[10] The style with which we respond to people and objects is, above all, the consequence of the structure of values that we have personally adopted. Hence there is no style without the capacity of the body to preserve all the values transmitted to us by a social group and acquired through significant experiences.[11]

Aesthetic experience of movement | We generate, and respond to, rhythm with our whole being: not only with movements, but also with our emotions. I suspect that, in public places or family circles, people are drawn into a "conversational waltz" because of the pleasurable experience that rhythm induces. Some movements, as most of us know, produce an exhilarating and stimulating feeling. This is not the same state of ecstasy, euphoria, or intoxication in which we might find ourselves while taking part in certain rituals. Expressions such as enchantment, delight, rapture, captivation, excitement, and inspiration seem to more appropriately describe the felt sensations. Oliver Sacks, the well-known writer in the field of clinical neurology, described the unbounded joy he found while walking, thoughtlessly and spontaneously, across a room. His satisfaction derived from the sudden feeling of unity with the "walking's natural, unconscious rhythm and melody," from the awareness of his body that "became music, incarnate solid music."[12]

The joy of sheer doing – its beauty, its simplicity – was a revelation: it was the easiest, most natural thing in the world – and yet beyond the most complex of calculations and programs. Here, in doing, one achieved certainty with one swoop, by a grace which bypassed the most complex mathematics, or perhaps embedded and then transcended them. Now, simply, everything felt right, everything was right, with no effort, but with an integral sense of ease – and delight.[13]

Sacks, giving the reader an eloquent account of his perception of "kinetic beauty" or "musicality of motion," shows that, under some circumstances, even the simple motion of walking is able to elicit a deeply fulfilling aesthetic experience.

To Sacks, it was the sense of organization and instant co-ordination of the movement that conveyed the feeling of "heavenly ease." He suddenly realized that, without conscious calculation, he was able to give himself to the "activity's own tempo, pulsion and rhythm." Yet such a feeling of getting into the rhythm did not at all take away the awareness of walking with style – a style that was "inimitably my own," as he put it. The legs felt "alive, and real, and mine." The sources of his aesthetic delight were both the awareness of a rhythmic order and the opportunity to consider the movement truly his own.

Following the illuminating analysis of Sacks, I would like to further focus my attention on the role that the rhythmically tuned body plays in an aesthetic motor experience. What gives the movement an aesthetic value? What are the qualities and determinants of the motor behaviour that induce a sense of contentment? The various authors, each being inspired by a particular philosophical option, advance diverse answers to these questions.[14]

Some consider beauty or grace to be the primary characteristic of the movement endowed with an aesthetic value.[15] A movement is beautiful or graceful when an idea, an intention, a meaning, an excellence, an inner unity and wholeness, or something "transcendent" and "inexhaustible," becomes manifest in a sensuous and dynamic form. Our aesthetic experience consists of the perception of irreducible excess, superabundance, and plenitude in a technically flawless motor performance. Technical perfection consists, above all, of harmonic order and inner unity.

Others prefer to pay attention to the formal qualities of motor behaviour.[16] Here the aesthetic value is in the successful realization of previously identified criteria such as rhythmic order, regularity, symmetry, balance, proportion, precision, harmony, versatility, surprise, and difficulty. Even though the motor form is not subordinated to external and pragmatic goals, it nevertheless remains bound to some "immanent laws" and principles. Empirical observations allow us to analyze and compare these principles and to recognize their communicative significance. Our aesthetic enjoyment springs from the perception of a correspondence between subjective performance abilities, and fixed, standardized movement possibilities.

The third approach considers movements from a subjective point of view.[17] Here the aesthetic is not merely a matter of adapting movements to objective qualities, but derives from the production of a dynamic form that on the one hand expresses ideas, conceptions, emotions, fantasies, and on the other elicits an awareness of total bodily involvement. To relate personal meanings to movements means to go beyond the factual, efficient, and useful and to place the movement in a context where expression is valued over performance. The deployment of symbolic figures and illusory appearances produces an aesthetic delight and, consequently, sustains or reshapes feelings. Both the figurative transformation and the refinement of feelings happen without adaptation to a conscious purpose; they are spontaneous processes since they originate in the primary need for the embodiment of inner life and from the "impulse toward symbolic formulation."[18]

Let us briefly consider the characteristics of movement within the third perspective, because it places great emphasis on the body's creative abilities and the affective component of the aesthetic perception. The aesthetic delight does not depend on the physiological or muscular processes alone, but rather on how we perceive ourselves in relation to the movement and a particular space. We experience a feeling of lightness and ease as we move with unusual dexterity and alertness and trust our own bodily capabilities. We do not perceive the swimming pool or the playing-field as a space to confront and conquer, but as a support and source of the body's dynamic impulses.

A fundamental prerequisite of aesthetic satisfaction is our ability to smoothly and correctly co-ordinate a great number of partial movements. The movement must exhibit an order, a structure in which the different segments obtain their unity and cohesion. When an adequate mastery of certain techniques is not acquired, the various elements follow each other without accentuation, articulation, or synchronization, making the movement devoid of internal coherence or "kinetic melody," to quote Paul Guillaume.[19]

However important it is, the rhythmic "melodic flow" alone is insufficient to produce an aesthetic value. What is needed is authenticity and expressiveness. The aesthetic enjoyment also arises from the expression of a momentary mood (of celebration), thought (of thankfulness), or desire (of stability) through original and harmonious movements. These achievements can neither be brought into existence on command nor narrowed down to stabilized and measurable patterns. They occur and develop, without any conscious planning and control, through the unconcerned variation of the symbolic structure, the playful improvisation of a kinetic theme, and the qualitative use of motor options.

RHYTHMICALLY ORGANIZED MOVEMENT | What are the organizing factors that endow a motor form with a desired unity and cohesion? How do we bring together the different segments of the movement and succeed in producing a harmonious form? But first, how does our body understand and play the "kinetic melody"? Primary motor abilities (endurance, strength, speed, flexibility, agility, and skill) and proper execution of the movement are essential elements of a harmonious form. We understand a movement when we perceive and feel it as an articulated and dynamic form that comprises some dominant elements. "To understand the movement," says Paul Guillaume, "is to organize its perception."[20] In fact, during the period of learning, it is our perception that guides our movement – the perception of a global form, not a detailed one. When, for instance, we learn to tie our shoelace, our movement is organized according to the salient moments of the schema. At the same time, the articulated outline becomes a dynamic structure, a virtual movement, and an anticipation of the way we will co-ordinate the main elements of the movement. We discern this kind of dynamic characteristic when we try to master a difficult movement pattern (turning while skiing); it intervenes as a facilitating and guiding link between our intention to move and the actual motor performance.[21]

Even if we successfully internalize the "kinetic melody," we are still unable to reproduce it as a harmonious whole without some familiarity with the material and spatial characteristics of the situation. It is essential, at least during the period of learning, that we experience the resistance of water, snow, or turf. We cannot, it is evident, learn to swim outside water or to ski without gliding on the snow surface.

Beyond the necessary motor abilities and the guiding perception of form, a movement receives its coherence and organization from the body's propensity to apprehend and produce rhythmic patterns.[22] Rhythm is the pivotal shaping factor that co-ordinates the movement's temporal segments into a harmonious form. When a rhythmic order is in place, the movement sequences are perceived as a unified and controlled reality and we then find delight and sureness in the elementary experience of accord between our intention and the actual motor performance: we can do what we proposed to do.

To move rhythmically means to repeat similar movements or motor elements. Such a repetition occurs when, for example, we swim breaststroke. Here an accelerating pull (outwards and backwards) of the arms is followed by their forward push, and a powerful frog-like kick of the legs alternates with their recovery through the bending of the knees. We repeat not merely a particular form of movement, but what Paul Souriau calls a "real rhythmic phrase."[23] This phrase usually comprises three elements: preparation, accent, and echo. The discus throw, or the basic parallel turn in skiing, illustrates quite well how the preparatory and echoing movements complement the decisive principal phase, and together constitute, as it were, a rhythm within a rhythm.

The elements of a rhythmic sequence are in a reciprocal and complementary relationship. Continuous movements such as rowing or skiing require the successful grouping of these elements. The overall meaning of the movement determines the value and function of each of its components. The specific articulation of the parts not only organizes a movement in time but also endows the form with an unmistakable character. Variations in speed affect our experience of the movement's intensity and quality.

Another fundamental feature of the rhythmic form is the emphasis or accent placed on certain motor components. It is the accent that endows the movement with a subjective character. By establishing a qualitative difference between accented and unaccented parts, we perceive the rhythmic structure as a "product" or "extension" of our bodily capabilities, namely our sense of rhythm, and not merely as the outcome of our passive and mechanical adaptation to a series of uniform pulses. We are captivated! We rejoice at being the author of our global motor experience. Beyond the sheer pleasure of reaching a goal through the movement and the already mentioned experience of accord between the idea or desire and its realization, it is the awareness of our body's dynamic possibilities, and of our subjective way of using them, that awakens a singular contentment in us. Thus, as we actively separate motor components, the movement becomes not only more harmonious and precise, but also emotionally appropriated. Because of our emotional identification with the focal points of the rhythm, the movement holds us in its spell and invites us to uphold its dynamic flow.

Some, however, place less emphasis on the body's potential to generate rhythm and contend that conscious intention presides over the ordered emergence of motor forms. The body's natural rhythms are merely "materials" that have to be consciously recognized and modified. To them a specific temporal organization and its variation are not the outcome of organic processes. They require a "will to form," an "inner activity," a conscious control over the process of co-ordinating the motor segments. "The rhythm of a movement," as Peter Röthig sums up, "has its centre in the psychological-mental experiential sphere."[24]

In numerous situations, the conscious ordering of motor elements is indeed required. However, there are instances when our body's endogenous sense of rhythm plays a much greater role in the articulation and accentuation of movement than do conscious planning and controlling.

Due to their relative independence from particular goals and directions, expressive movements do not seem to require a concentration on numbers and measures. Their frame of reference is neither a specific distance nor a location and limit, but the movement as it is related to a friendly context or a pleasantly absorbing challenge. In the temporal structuring of movement it is more important to experience the spatial and material qualities than it is to be aware of our voluntary intentions and efforts.

A handicraft activity could be one of these instances. Because of the intense sensorimotor involvement in the making of furniture, ceramics, or musical instruments, craftsmanship is considered to be a highly rewarding human activity. Lewis Mumford pointed out that one of the beneficial effects of craftsmanship is the intensification of the body's "natural organic processes."[25] The violinmaker, stone mason, and ceramist set and adjust the hand's rhythm according to the properties of the materials dealt with. While, at the same time, obeying the natural rhythm of their hands, they experiment, try out different solutions, and, above all, enjoy the "privilege of handling" (*privilège de manier*) wood, clay, or stone.[26]

DANCE | Dance also illustrates how the "nondirected and nonlimited" movements mobilize some of the indwelling capabilities of the body. I have mentioned, in an earlier paragraph, Straus's analysis of the complex and subtle relationship between movement and space. "Expressive movement cannot be produced apart from the immediate experience of which it forms an integral part. The immediate experience and the movement in which it actualizes its meaning are indivisible."[27] In dancing, Straus tells us, we do not move "through" space, from one point to another; we move "within" a space, where we are no longer guided by a system of axes and directions. Our movements are responsive to a spatial structure, and here the topical preferences have been abolished. In such a "homogenized space," we neither seek to reach practical goals nor produce any change, but merely enter into it as participants and surrender ourselves to an activity freed of direction and limit. Because the dance itself has its own intrinsic value and is determined by the symbolic and non-practical qualities of space, the movements are performed with facility and delight. Forward or backward, the movement is carried out with equal ease. Whereas, in our practical life, the turning and backward movements are disagreeable and provoke discomforts (dizziness and fear), in dance, their unfolding yields to the pleasant sensation of rapture and sometimes even of ecstasy.[28]

We experience our whole body in a similar manner: we perceive it with a sense of unity and not as an object that we have to guide and control. What fosters such an inner consciousness of the body is the transposition of the ego or "I" relative to our unreflective bodily schema. In the phenomenological sense, the "I" of the active person is located somewhere in the region of the eyes. Since the trunk's activity becomes dominant in dancing, our "I" moves from the eyes to the trunk. The purposive consciousness pulls out, so to speak, and allows the abilities of lived body to form the movement. "The crescendo of motor activity in the trunk accentuates the functions of our vital being at the expense of those which serve knowledge and practical action."[29]

Rudolf Arnheim agrees with the claim that such a shift from the head to the torso triggers the temporary leave from conscious control in favour of the spontaneous and instinctive impulses. Dance requires the giving up of the "safe control of reason and modesty" and the trustful surrender to the vitality of the body. "This paganism of dance accounts for its wholesome therapeutic effect on emotionally inhibited people."[30] We surely experience some restorative pleasure of self-expression while we skilfully control our rotating movements. But to feel a truly healing terpsichorean effect, we have to move without the guidance of our purposive consciousness.

The expressive movement that leads to an aesthetic enjoyment, and dance, are similar in many respects. Yet the former is broader in scope than the latter. Specific cultural principles, qualities, and costumes shape a particular dance. They provide its grammar. According to Judith Lynne Hanna "a grammar (syntax) of a dance language, a socially shared means for expressing ideas and emotions, is a set of rules specifying the manner in which movement can be meaningfully combined."[31] Expressive movements are generally neither planned nor produced on command; they occur, unexpectedly, when the body is allowed to exhibit its tendency to produce exploratory and non-functional motions and express, through this rich and surprising spectrum of movement compositions, momentary feelings, ideas, and fantasies.

As I have said before, both dance and expressive movements are "out-of-the-ordinary" activities, in which the adequate rhythmic structuring plays an important role. Time and again, the movements themselves suggest or dictate a specific rhythmic order. How does this happen? The movement, as previously stated, is made up of sequences and phases that are connected together by "points of junction" (*Knotenpunkte*), to cite Arnold Gehlen.[32] The whole movement is held together by these "joints," because, in a sense, it is only through their successful co-ordination that the correct execution becomes possible. Arnold Gehlen speaks of the "symbolic structure of the movement" because these pivotal points not only hold together, but also represent, the entire movement phrase. For instance, when we attempt to execute a difficult motor combination, we merely have to focus on these "crucial moments" and come to accomplish, automatically, the so-called in-between phases.

More important perhaps, due to their "points of junction," movement phrases themselves suggest a particular temporal configuration. There will be a variation of all subsequent rhythmic patterns even if we slightly change our manner of executing these fertile elements. Here the change pertains to the intensity of an accent and not so much to the emphasis within a movement sequence. A stronger accent placed on the pivotal point of a turning movement could easily affect its outcome; it could, for instance, prompt our body to execute a jump rather than another step or turn. When we alter the tempo of our stride or switch from walking to skipping, the rhythmic configuration of the movement derives also from the particular way we place an accent on some "points of junction." The unique "tensional quality" of the movement itself is the determining factor, not the conscious representation. We allow our body, in Ursula Fritsch's telling words, "to think by means of the movement."[33]

Maxine Sheets-Johnson, in her analysis of the nature of rhythm in dance, also believes that the "dynamic line" of the dance movement itself suggests a specific temporal flow: "Because time is not a thing which preexists and awaits carving up by the dancer, because it is something created by the dance itself, it exists specifically only in relation to a specific movement within the dance."[34] Merce Cunningham put forward a similar opinion: "You have to get the idea that movement comes from something, not from something expressive but from some momentum or energy, and it has to be clear in order for the next movement to happen. Unless you can begin to see that way, you don't get a progression in the movement, a going from one movement to the other, which seems logical. By logic I don't mean reasoning but a logic of movement."[35] Each movement has, as both Sheet-Johnson and Cunningham have pointed out, an intensity and force and these either induce reinforcement or prepare a qualitative change – from weak to vigorous, from gentle to aggressive, or from contractible to expansive. Moving according to a kinetic logic does not consist of repeating familiar rhythmic patterns but of allowing the unfolding of rhythmic structures to be dictated by the dynamic flow of the movement itself.

Reflecting on how the movement operates, how a dancer gets from one movement to the next, Merce Cunningham introduced another term: *eloquence*. This term refers to the movements' expressive power – expressive in the sense that a movement "wants" to unfold in a certain way, independently of the dancer's conscious desire to articulate an emotion or a meaning. Endowed with thrust and force, movements do not merely "seek" to represent something but also indicate, show, and project possible qualitative changes – especially a change in their rhythmic configuration.

In a different context, Arnold Gehlen rightly considers the ability to execute and co-ordinate "intelligent movements" as one of the most important features of human life.[36] Human movements are intelligent because, on the one hand, the subtle variations of the rhythmic pattern, tempo, form, co-ordination, and function can be produced, without any conscious representation, as motor responses to the movements themselves. On the other hand, these bodily achievements develop in parallel to some thought processes such as counting, restarting the count, combining previously unrelated ideas, or suggesting solutions. Intelligent movements are neither the results nor the conditions of thought processes. Thought and movement, nevertheless, seem to become more articulate and refined in conjunction and thus exert a beneficial influence on each other.

The execution of polyrhythmic movements is also made possible by the body's capacity to act in an intelligent manner.[37] Concentrate on the hands of a virtuoso and pay no attention whatsoever to the sounds, advises Paul Valéry.[38] Do this and you might come to view the hands as dancers who follow two sorts of rhythmic order. Move your arms and legs according to different rhythmic patterns and produce similar counter-rhythms. Here again, the source of this complex motor performance seems to be the body's ability to generate muscular impulses and not the representation of a rhythmic scheme.

surrender to the body | Dance, music-making, and aesthetic movement experiences imply an attitude that may be called *renunciation*: a relaxed and trustful surrender to our bodily impulses and intentions. The movements are not only upshots of specific intentions, but also responses arising from the formative powers of our body. As we move easily and effortlessly, we abandon ourselves to the body's sense of rhythm that, without purposeful pre-assessment or planning, introduces new patterns, and responds appropriately to the demands of the motor situation. Merce Cunningham stresses the importance of confidently relying on the resources of the body. Dance, as he put it, is "the play of bodies in space – and time" and not "the product of *my* will." "But the feeling I have when I compose in this way is that I am in touch with a natural resource far greater than my own personal inventiveness could ever be, much more universally human than the particular habits of my own practice, and organically rising out of common pools of motor impulses."[39]

When Cunningham invents surprising and unusual rhythmic patterns, he allows his body to "think by means of movement," remain attentive, open to the suggestions of the movement.

This bodily potential, through which a "strangely spontaneous and strangely contrived" rhythmic organization occurs, is also brought to our attention by Paul Valéry. The dancing body "assumes a fairly simple periodicity that seems to maintain itself automatically; it seems endowed with a superior elasticity which retrieves the impulse of every movement and at once renews it. One is reminded of a top, standing on its point and reacting so sensitively to the slightest shock."[40] Elsewhere, speaking of the temporality of dance, Valéry expresses the same idea:

By Time I mean organic time, such as exists in the ordering of all the alternating and fundamental functions of life. Each of these is affected by a series of muscular acts which reproduces itself, as if the end or fulfilment of each series brought about the beginning of the next. On this pattern, our limbs can carry out a set of figures that are all interlinked, and whose repetition brings about a kind of exhilaration, ranging from languor to delirium, from a sort of hypnotic abandonment to a sort of frenzy. In this way the condition of dancing (état de dance) *is created.*[41]

What is this *état de dance*? It is a state that we could call, following the insights of Straus and Buytendijk, pathic.[42] While dancing, it is the rhythmic series of forms that energize, stimulate, and even compel the dancers to execute movements. Like the musical tones, the "eloquence" of movements affects them, draws them further into the experience and, at the same time, evokes in them a feeling of freedom and power. It produces the rhythm and triggers its variation. At the same time, it gives us the sense of being carried along. "For rhythm," says Carl E. Seashore, "is never rhythm unless one feels that he himself is acting it, or, what may seem contradictory, that he is even carried by his own action."[43] While walking, swimming, or dancing, we all have been captivated by our ability to produce rhythm; the feeling of elation, soaring, and ease, the disappearance of tensions and confrontations, prompted us to go along and further a pleasant experience. We perceived in the rhythm a playful interaction between a call from, and a response to, a motor event.[44]

Such a rhythmizing experience is suitable to convey a feeling of time – a feeling of personally adopted temporal order.[45] The more we execute and practice rhythmic figures – in dance, gymnastics, or music-making – the more we feel at ease to introduce variations in the tempo of an activity. We develop in ourselves the sense of intensity. Rhythmical exercises help us to become aware of all sorts of temporal ordering and thus to successfully manage the various tasks of our daily life.

Doubtless, a movement impulse is not the only determining factor for the rhythmic configuration. Musical sounds or atmospheric impressions sometimes exert a more formative influence on us than the movement itself. The degree of rhythmic variation of our swimming strokes significantly depends on the characteristics of the concrete aquatic area: every swimmer knows that an agitated sea elicits very different motions than a calm lake.

Rhythm may also spring from the subtle interplay of muscular contraction and release, tension and relaxation.[46] The body then expresses its own pulsating life, recurrently placing accents on certain parts of the movement. Thus Paul Souriau explains motor rhythmicity by the "law of compensation."[47] In numerous forms of bodily performance, an expenditure of energy is followed by the sensation of fatigue and the need of recuperation. The interplay of periods of intense effort and compensatory calm generate the alternation of strong and weak beats. Such a rhythmic unfolding of an activity cannot occur without the effort necessary for its execution. Movements have their cost and must be balanced by an appropriate recovery.

Certain feelings such as embarrassment, boredom, or timidity easily yield to rhythmic movements. They serve to free us from an unwelcome situation. The urge to move is closely bound to the need to find release from the weight of a particular feeling. Because the movements themselves exert an attraction on us, we start to play with them and, in this manner, make an attempt to overcome our unpleasant feelings.[48] Thus rhythmic movements are able to "correct" an emotional state or elicit further feelings, particularly the agreeable "feeling of being alive."[49] It is for this reason perhaps that physical education and sport could play a very positive role in our lives. If sport is freed from the exclusively utilitarian and rationalized perspective, it could lead to the elementary "enjoyment of existence" (Buytendijk), to an alert receptivity to our bodily resonances.[50]

6 | Memory

The body as a temporal form | Merleau-Ponty considered the moving body as a "medium of our communication with time as well as with space."[1] Gabriel Marcel expressed the same idea when he defined the body as a "temporal form."[2] Indeed, our past experiences, the painful as well as the pleasant ones, are inscribed in the body. If we carefully observe the actor's or athlete's behaviour, we will catch sight of a particular history and, at the same time, of some individual possibilities. Arising from the body's natural dispositions and powers, these possibilities expand or shrink with the ongoing acquisition and modification of experiences and interests. The body is a "provisional sketch" of our "total being" inasmuch as it is both "rooted in nature" and "transformed by cultural influences."[3] It is always our past and future, always something already formed and being formed.

The structure of action-time is not consciously represented, but lived as an intrinsic factor of our motor experience. Whenever we act, time is constituted of the chosen or imposed possibilities, acquired experiences, and actual execution of movements. These three temporal dimensions determine and generate each other so that each one receives its character and significance from the other two. The possibility of my jumping over the bar or returning the tennis ball is determined by the skill level attained during past practices. Conversely, I am able to acquire, develop, and perfect a skill if I am exposed to increasingly difficult challenges. The actual moment of my jump or forehand stroke is defined by my previous experiences and concrete possibilities pertaining to a given situation.[4]

Paul Ricoeur pointed out that learning is essential to all habits and the process of acquiring habits brings into the fore one of the pivotal aspects of human life: time. "The key idea of habit ... is that the living being 'learns' in time. To reflect on habit always means to refer to the time of life, to the holds which a living being offers to time and the holds which, thanks to time, he acquires on his body and 'through' it on things."[5] I would think, however, that rather than offering holds to time on our "timeless" body, the learning of habits makes possible the display and articulation of the body's inherent temporality.

In possession of its already formed dispositions and skills, the body manifests its availability: it is able to do something. Though we seldom stop and rejoice about its fitness and usefulness, the available body announces itself in almost every moment of our active life. "The availability of the body is experienced without thinking in the execution of habitual actions, but this experience remains nevertheless continuously in effect. It manifests itself in the certainty we feel about our ability."[6] In other words, we find the body available while we face a particular task such as playing a difficult rhythmic passage on the piano or climbing a steep rocky slope. The availability is not based merely on the body's anatomical and physiological systems, but also on the acquired skills. We are able to flawlessly co-ordinate the elements of a difficult movement because we have done it before and are able to remember how we did it. If we happen to venture further and propose new motor forms, it is still the secured skill that makes possible the execution of an inventive solution. Our body develops skills and is able to deal with new situations, and these two aspects, skilled knowledge and creative availability, develop in reciprocal interaction. An acquired skill sustains and increases the body's availability and the availability that we inventively use, in its turn, perfects the skilled behaviour. Thus our body relates to what is "behind" itself and what is "ahead" of itself; because it is invested with a memory, it forgets nothing, and also with a remarkable sense of the possible, it anticipates habitual or probing movements.[7] There cannot be anticipation without the body's ability to store past experiences and convert them into skills and, conversely, no retention can occur without anticipating habitual or new motor accomplishments. Although no strict separation is possible, in this chapter I focus on the memory of the body.

SKILL AND HABIT | Skill and habit are so closely related that it is not easy to provide a distinct definition of the terms. *Skill* often refers to a capacity to perform some actions. The concept of *habit* evokes both the acquisition and the actual and appropriate use of a capacity. Habits may be launched in diverse situations; skills tend be more specific. Although both can be spontaneous and dextrous, habit brings to mind the notion of ease, skill that of proficiency. There are some habitual actions that do not require skill (going for a walk) and, inversely, skilful performances that, though not devoid of efficacy, lack the fluency and facility of habit (my playing on the piano).

Skills are acquired through our implicit memory or, as Jerome S. Bruner would put it, a "memory without record." It makes possible the conservation of perceptions and movements that have occurred in the past and that have led to the learning of new ways of moving. As its name indicates, we learn to swim, ski, or ride a bicycle without being able to recall particular learning events or experiences. These are "converted into some process that changes the nature of an organism, changes his skills, or changes the rules by which he operates but which are virtually inaccessible in memory as specific encounters."[8] Skilled behaviour is defined as an aptitude to perform movements learned through successive experiences and leading to the proper understanding of the movement patterns. The practice of motions transmute into a "latent knowledge" (Merleau-Ponty) of our body, and the "sedimentation" takes place without "keeping record" of each step of our progression.

Merleau-Ponty and Buytendijk tell us that we do not acquire habitual movements by merely creating a series of associations between impressions and motor impulses.[9] We actively adapt ourselves to specific situations through the reorganization of our movements, these being guided by the perception of a meaning, and by the awareness of our possibilities.

When we learn the piano, we execute particular "*modes* of movement" (Straus) according to the sounds we hear and expect to hear, and the music's emotional and aesthetic qualities.[10] We learn the speed, direction, strength, and effect of the movement in relation to acoustical and aesthetic configurations. The repetition of the appropriate motor response leads to the stabilization and articulation of the form, the widening of our motor possibilities, and the development of skills. Connections between perceptual content and movement response are established in light of the meaning, motive, or goal understood by the "knowing-body." "The acquisition of a habit is indeed the grasping of a significance, but it is the motor grasping of a motor significance."[11] Our body knows, or pretends to know, how to lift a seemingly heavy object or to walk on slippery ground, and it does so in an unreflective but intelligent manner.

In his definition of skill, Bruner also emphasizes the body's relation to an overall meaning and to some future development: "A skill is a mode of sensory-motor functioning that provides rules for anticipating and responding to categorized situations of varying uncertainty."[12] There is nothing mechanical or automatic in the production of the skilful action since we actively apprehend a meaning, and the action-situation continuously changes. Occasionally, our body "emancipates" (Buytendijk) itself from the already acquired rules, and responds to the requirements of a situation with a new movement. The new skill swerves from a past experience or a previously applied schematic response. Yet the acquisition of the new is only possible because some elements of the innovative variation are contained in the old structure. When, for instance, my son reaches for a frisbee by turning sideways, bringing his left shoulder to the front, and placing his open right hand behind his back, he improvises a new motor performance on the basis of an already acquired dexterity, which comprises the discernment of direction, transition, and limit of the catching movement.[13]

Once a movement is learned, it is available for repeated and flexible use. It functions "in the manner of the organs" (Ricoeur), it becomes a habit. Paul Ricoeur rightly considers the "use-value" one of the essential characteristics of the habit: we know how to do something, we can do it. "Habit ... is a power, a capacity to resolve a certain type of problem according to an available schema: I can play the piano, I know how to swim."[14] The schema is a spatial-temporal one: it is a structure that allows us to move in our surroundings with ease and confidence, because it consists of the correct evaluation of the distances between our body and the objects and the corresponding temporal relations.[15] Our hand knows how to reach for, and use, a doorknob, a lamp, or a waste-paper basket. Before catching the ball, it senses the appropriate spatial position and the time it needs for the right positioning. If, for some reason, someone modifies the location or functioning of these objects, we immediately take notice of the acquired habit of our body.

As a result of the repeated communications with objects, a reliable spatio-temporal schema – the bodily knowledge of a room, a garden, a swimming pool, or a street – does not consist in an objective representation, but in a "certain modulation of motricity." "My flat is, for me," affirms Merleau-Ponty, "not a set of closely associated images. It remains a familiar domain round about me only as long as I still have 'in my hands' or 'in my legs' the main distances and directions involved, and as long as from my body intentional threads run out towards it."[16] Perhaps no other human activity contributes so much to the building of these dynamic "threads" or schemata as play. It is obvious that, while playing, children progressively develop skills and conserve information about a wide range of available objects. To acquire a spatial-temporal schema, their sense of vision is insufficient.[17] In order to succeed in adjusting motor activity to spatial characteristics, they have to rely on their sense of touch. Above all, it is the tactile exploration of space that develops the trustworthy awareness and keen appreciation of directions and distances.

To have proficiency, an unreflective relation to the world is invaluable. Samuel Butler accurately stated, in his remarkable book, that the older our habits are, the less control and reflection we need in order to carry them out.[18] Reflection and attention can even be harmful since they yield uncertainty and hesitation. Lack of control and consciousness, on the other hand, leads to the achievement of a harmonious and efficient performance. Hence the paradox: when we know something, we do not know it, in the sense that we are unconscious of it. We truly know how to sail a boat or play an instrument when we are not conscious of our knowledge. This law holds not only in respect to our bodily habits, but also to our modes of thinking and the making of value judgments in general. These, after long practice, function almost unconsciously, just like the movement of our tongue does when we eat or speak.[19]

For the execution of skilful actions, we do not need to consider and bring together the various movement segments: the movements, even the difficult and complicated ones, are performed as a whole. They are fluent, accurate, and speedy if we do not attend to the specific elements of the global form. Uncertainty and hesitation stem precisely from the attention we pay to them. Musicians declare that they are able to play a piece well when they do not consciously strike every single note. Some would say: "The piece is in my fingers." During the process of learning, the consciousness of the details forces them to play slowly. When they know a piece, they no longer know it.[20]

Neither the playing hands nor the instruments are objective realities that musicians consider analytically. An organist, to borrow Merleau-Ponty's example, does not analyze and form an objective representation of the new instrument that he is going to use. He relates to the distances and directions on the basis of his readiness to play the music and communicate its emotional content. "During the rehearsal, as during the performance, the stops, pedals, and manuals are given to him as nothing more than possibilities of achieving certain emotional or musical values, and their positions are simply the places through which this value appears in the world."[21] His reading of the score and the sounds produced trigger and define the orientation of his movement in relation to the immediate space. His expressive movements, perfected through many hours of practice, are the means used to create a link between the written notes and the actual sounds. His body exists as a "mediator" between his intention to express the composer's ideas and emotions, and the outcome of his play on the instrument, the music itself. The acquired habit creates music and conveys an artistic meaning, just as much as the awareness of the dynamics and possibilities of tones enables the artist to introduce fine variations.

INVENTIVE STYLE | Acquired habits allow us to extend the limits of our bodily existence. Objects – a car, a telephone, or a pencil – become the means by which we create relations and open ourselves to new experiences. "Habit expresses our power of dilating our being-in-the-world, or changing our existence by appropriating fresh instruments."[22]

As with our hands or legs, these objects are experienced as silent but familiar mediums, opening up for us various possibilities. Without thinking, we immediately understand how to use them in order to reach our objectives.

When our body has a "knowledge of familiarity" in the presence of some specific objects, it performs the appropriate motor response. The visual or tactile perception of a jar or a doorknob is immediately followed by an accurate and harmonious execution of the movement of turning. Merleau-Ponty observes that the perception induces a "certain style of motor responses," because each object suggests its "motor essence."[23] Ricoeur also speaks of a movement that "becomes a whole, a stylized harmonious *form* which adheres to the perceived form without either a guiding image or a special initiating order."[24] The concept of style refers here to an efficient execution pertaining to a particular action-situation.

Style is not only the characteristic feature of expressive realities (a melody or a painting), but also that of perception itself. Our vision, bringing together the manifold impressions and aspects of the world in light of a global meaning, exhibits a style. But the coherent perceptual point of view is closely linked to a distinctive way of moving, which also reveals a style. "To learn to see colours it is to acquire a certain style of seeing, a new use of one's own body."[25] Both perceptual and motor activities unfold due to some interpretations that are ascertained by the body. Hence the body is not merely an object of our thought, but a dynamic series of performances that carry a "cluster of meanings." It can be compared to a work of art; it is a "nexus of living meanings."[26] When perceiving and moving towards something, the various bodily parts perform a coherent, stylized action because they are all involved, and linked together, with the same lived meaning.

We daily experience an ongoing modification of the meanings of objects, to which a reference has already been made. A gift received from a beloved person differs in appearance from the very same object observed in the store. The "look" of the much cherished house changes significantly once we learn that it is to be sold or demolished.[27] The change is triggered, in part at least, in accordance with the experiences, feelings, values, and expectations preserved by our body.[28] The body of any ball player has assimilated, in addition to skills and experiences, the implicit obligation of fair play. The meaning of the overall game situation is evaluated according to the tacit knowledge of rules, obligations, and possibilities. The body confronts a defensive or offensive task not only with its already acquired skills, but also with a consciousness of what is desirable, possible, and permitted. The same wisdom governs, to a large extent, the movements and perceptions of the car driver. Here, too, the hands and legs have incorporated a large number of norms and values, and the meaning of what the driver sees and hears is given through the implicit awareness of possibilities and interdictions. The combination of conscious intention and the efficient use of an anonymous, latent knowledge of the body results in the ability to drive a car. When we drive, "we merge into this body which is better informed than we are about the world, and about the motives we have and the means at our disposal for synthesizing it."[29] It is our body that understands the surroundings (road, traffic lights, other cars) as a meaningful whole and actively regulates the movements on the basis of an incorporated knowledge, in which the "feeling of responsibility" is paramount.[30]

The acquired style does more than merely helping us to execute, naturally and with ease, some repetitive actions and, thus, to deal with the requirements of a task. Our handwriting does not consist only in the execution of a series of purely mechanical movements; it is, above all, "a general motor power of formulation capable of the transpositions."[31] The structure of our movements is variable according to the characteristics of the circumstances, especially when we handle familiar objects such as a hammer or a lighter. Unlike some devices that call for rigid, stereotyped gestures, the objects occupying our living quarters may suggest new ways of using them; they demand "intelligent handling" (Ricoeur).[32] Occasionally, when we are not concerned with the achievement of a specific result, we savour the modification of movement for its own sake. It is the hand's feeling of the possible that discovers a new way of using a tool or an instrument.

Beyond the variation of movements, the style also gives rise to surprising innovations. Stendhal's writing style, his "system of speaking," constituted through long practice and observations, reveals his character and perception of the world and also "allows him to improvise."[33] In a similar fashion, during the execution of habitual gestures, our body is able to invent new solutions and thus transgress the familiar.

All the monographs on acquisitions of habits point to this curious relation between the intention which launches out in a specific direction and the response, arising from the body and the mind, which always has the air of an improvisation. This is familiar in the case of skaters, pianists, and even aspiring writers. Habit only grows through this type of germination and inventiveness concealed within it. To acquire a habit does not mean to repeat and consolidate but to invent, to progress.[34]

Again, inventiveness is one of the principal features of children's play. Ritual play consists, in part, of reproducing, more or less faithfully, some guiding models and of repeating a certain number of chosen conducts.

True, some make use of this activity for compulsive purposes: their movements are rigid and invariable. But it would be a mistake to conclude that all repetitive behaviour points to the child's deep-seated desire to find protection from life's uncertainties. Erik H. Erikson defines ritual play as a "mixture of formality and improvisation," a "rhyming in time."[35] These contribute to the child's capacity to enrich the motor form with new motives and elements, to deform it, even to change it to such an extent that it becomes a genuine caricature of the original behaviour.

If children relate to their surroundings with sympathetic ties and consider objects and people in a relaxed manner, they can avoid inertia, foreseeable and familiar movement patterns. They can propose new motor combinations to the same extent as musicians or dancers are able to introduce subtle rhythmic or expressive variations into their stylized performance.

THE GIFT OF AUTOMATISM | To be vivid, flexible and adventurous, habits must be educated. This is the advice of John Dewey who defined habits in terms of their inert persistence and not of their spontaneity. "What is necessary is that habits be formed which are more intelligent, more sensitively percipient, more informed with foresight, more aware of what they are about, more direct and sincere, more flexibly responsive than those now current. Then they will meet their own problems and propose their own improvements."[36] Notwithstanding our effort to make them inventive and flexible, our habits sometimes fall into automatism. We then have the impression that we do not entirely coincide with our body's abilities. They are, to a certain extent, distinct from us, resist our intentions, and reveal their unavailability. We tend to approach objects with an inflexible interest and provide a knee-jerk response to their demands.

Our blunders are often caused by an automatic release of movement: we are unable to catch a ball because we approach the task with the wrong type of movement. The same may occur in our speech as, under stress, we mispronounce a word. Perhaps more frequent are those gestures that we mechanically execute in situations presenting some modifications and requiring a readjustment of our behaviour. We inadvertently throw something on the floor because the wastebasket has been moved. Or we reach in vain for the gear shifter when we happen to drive a car with automatic shift. We simply fall back to our old habits. "Repetition of daily cycles of action saves the trouble of inventing. For reasons of economy we appeal secretly to old resources and yield to them."[37] Intermittent lack of attention also leads to the execution of automatic movements. This happens, for example, when our exhausted body continues to walk or swim and we no longer consciously control our movements.

Still, automatic movements can be useful when, as actively involved, we are required to make a quick assessment of a situation. Our perception is closely linked to our actual and virtual motor performances. While driving a car, the significance of the road, other cars and objects, or signs depends on whether our body is or is not able to execute driving movements.[38] The objects suggest, and evoke in us, previously experienced movements and their instantaneous repetitions. Because our body preserves motor experiences, we see dynamic qualities in the objects: we are able to assess their possible actions or reactions. We accurately evaluate how the rock, ice, or curve would "act" and accordingly, and without any delay, determine the motion of our car and the further development of our driving gestures. The movement leads to a perception of the dynamic characteristics of the object and these, in their turn, trigger the automatic execution of further movements.[39]

It would be, therefore, a mistake to belittle automatic motor performances. Some brain-injured individuals are unable to discriminate among possible choices according to a logical selection, but they still continue to perform the automatic, stereotyped movements that are significant to their life.[40] Hence the importance of an educational process that allows the body to incorporate and preserve movement patterns, as well as values. "Such patients," Tellenbach tells us, "can still radiate the atmosphere and exhibit the behavior patterns of the society in which they grew up. Without reflection or effort, they are able to do what 'one' does in a given situation. They can still largely grasp the meaning of a certain situation for themselves and for another person."[41] When the normal human existence is no longer possible, the body still presents possibilities for creating and entertaining a coherent relationship with fellow human beings. It succeeds even though the movements unfold according to a fixed scale of meanings and a limited sphere of interests.

7 | Imagination

MOTOR IMAGINATION | Our previous analysis of skilful and habitual actions drew our attention to the relationship between the body's capacity to preserve experiences and values, and to deal with new challenges. The act of touching illustrates the close connection between retention of past behaviour and anticipation of a particular movement. As our hand intends to pick up a glass or our foot moves toward the gas pedal of the car, we re-enact the past and expect a certain type of bodily contact with the object. The act of gently pushing the pedal or bringing the glass to our mouth involves both the anticipation and memory of a particular tactile sensation.

Just as our body remembers and anticipates sounds and odours, it is also able to recall and expect particular tactile sensations. The hand already knows what kind of impression that contacts with snow, iron, sand, or wood will produce. Similarly to the learning and refinement of any skill, tactile memories are acquired through the circular and reciprocal relationship between sensing and moving. In possession of these memories, surgeons anticipate both sensations and movements while making very precise incisions with their scalpel on the unseen parts of the body. My excellent dentist delicately uses his drill and relies on a similar tactile anticipation to direct the fine movements of his fingers. The basis of all deft motion is the body's ability to expect sensations and movements.

In order for the reader to better understand the continuum of bodily memory and anticipation, I would like to draw upon the original insights of Melchior Palágyi. In various publications, Palágyi discussed the most important characteristics of the sense of touch and motor or tactile imagination.[1] He demonstrated that, in touching, we not only experience, but also "imagine" specific qualities and forms. *Tactile images*, as he called them, and tactile impressions have equally important significance when handling objects. For example, reaching for the doorknob, our hand projects tactile sensations that correspond not only to the actual sensations, but also to our approaching movement. When someone's hand moves to touch a particular part of our body, we anticipate a yielding or resisting movement. Without such a bodily response, we would hardly notice the actual touching of our skin. We become aware of the imagined movement in as much as a very small movement of our body follows it.[2] Our bodily response would occur even in the absence of a tactile contact. Seeing a needle advancing to our finger, we experience some sensations before the object makes contact with our skin. We have an imagined sensation that constitutes a response to the imagined continuation of the movement.

Just as much as there is no actual tactile sensation without tactile anticipation or imagination, so there is no actual movement without virtual or imagined movement. Indeed, whenever we touch an object or step on something, our hands or legs anticipate the appropriate movements. Lifting up heavy objects requires a different movement than lifting light ones. The perception of heaviness prompts not only the actual motor approach to, and lifting of, the object, but also the virtual movement of approaching and lifting as well. We realize how intensely we anticipate a movement when, for instance, we are caught in a deception: the seemingly heavy object is, in fact, light.

Palágyi argues that the sense of touch is our most fundamental and complete sense since it allows us to distinguish between an object and its image copy. Our touch provides us with a primary experience of reality and space. It is the corresponding tactile imagination that elicits the perception of the location of the material object and its spatial dimension. The tactile contact of an object induces a virtual movement that leads to the experience of its spatial extension. "Imagination is, so to speak, the organ for the perception of space."[3] A common experience well illustrates the fundamental role that virtual movement plays in our knowledge of things. If we cover the rim of a cup with the palm of our hand, we touch it merely in two places. But the touch of two segments is sufficient to induce the imagined movement of the circle, and this virtual movement makes us perceive the cup's opening as round. Through virtual movements, we can correctly perceive the form and location of objects in space.

Not only tactile sensations, but also visual and auditory ones, provoke virtual movements. If we attentively look at a tree or a tire on a car, we touch these objects with imagined movements. As already mentioned, the basic affordance of a practical object or a work of art arouses in us the need to touch. The execution of virtual movements precedes the tactile contact and we feel this anticipation in our gently quivering hand. I would think that a petanque player, whose hand anticipates a throw that knocks away the other player's leading boule, feels an even stronger sensation. Once again, his virtual movement guides and determines the form of his actual movement.[4]

Imagined tactile movements are not to be equated and confused with the visual representation of movement. If such confusion occurs, it is due to the "tyranny" of the latter over the former. For us, as seeing persons, the tactile image is overpowered by our visual representation. Hence, most of the time, our imagined movements remain unconscious. Only in some particular situations may we become conscious of them, such as in a shopping centre or at the airport where we tend to use an automated sidewalk. Preoccupied with our destination, we usually step on this transport device without thinking. If, for some reason, the apparatus fails to function, we suddenly feel a queer sensation in our legs. The real sensation of the material, elicited by our actual movement, does not correspond to the anticipated sensation. Although we are well aware of the breakdown, our body's prediction remains a habitual one; it is appropriate for a mobile belt and not for an immobile one. When we use an escalator, our body ordinarily anticipates the movement and sensation pertaining to a moving surface (when we step on) or an immobile one (when we step off). Whenever, in our daily life, our hands or legs unexpectedly encounter a resistance, or conversely, its absence (reaching in vain for the last step on the stairs), we take some notice of our body's ability to imagine both movements and sensations that correspond to the real movements and sensations.

If we bring a piece of lemon close to our mouth, we anticipate a sensation, one that is induced by our imagined bite. True, the representation of a bite also prompts the feeling of an acid-like sensation.[5] If we anticipate the tactile contact with a metal door, our hand feels the cold of the material. Inasmuch as we anticipate sensations, which tend to trigger real sensations, we are able, in retrospect, to become aware of the imagined movement. In fact, we react to the actual sensation itself by performing virtual movements that summon up further sensations.

Motor anticipation is a unmediated vital process and not a conscious psychological event arising from a visual impression.[6] The born blind are able to imagine geometrical figures and anticipate a great variety of tactile sensations and movements and, therefore, their tactile imagination enjoys autonomy while, for the seeing persons, it is tied to visual representation. If I represent my moving palm around a ball, the movement and the ball are given together as a whole. For the blind persons, the round shape and the movement are envisaged as a dynamic process consisting of elements succeeding each other. The groping examination of their hand constructs the shape by remembering and anticipating both movements and sensations. These are structured not in terms of set directions, fixed goals, and measurable distances, but in those of speed, correlation, transition, and limit. They do not conceive a virtual movement as a unified form, made up of segments and held at a distance in space. For them, it consists of the awareness of an unfolding. We may illustrate the difference if we consider two different aspects of a throw: we might focus on the relationship between the throw and an external object, namely its destination, or on the feeling of the fluent progress of the motion that comprises possibilities (hit or miss). In the first case, we visualize a global form; in the second one, we consider a dynamic process, developing with an open outcome.

In fact, to execute a movement is to imitate the imagined performance. Unless the motor situation suddenly changes, the actual movement follows the anticipated one. The movement of turning the knob on a door is preceded by the turning motion that we execute in imagination. Whenever we envisage a dive into the water, our take-off involves an imagined movement followed by the execution of the dive itself. To state again, both the imagined and actual movements are in reciprocal relationship with the anticipated and actual sensation. There is, however, an important difference between the two kinds of movement. The imagined one does not encounter material resistance. It lacks the kinaesthetic experience of tension, relaxation, pressure, or the shift of weight of the body – all indispensable information, required for the proper learning and control of movements.

Still, motor skill proficiency calls for the "schooling" (*Ausbildung*) of motor imagination.[7] It must become keener and more focused. But this, on its turn, can be perfected through the execution of tactile movements. Palágyi suggests that we first press a small rod against our free hand and, subsequently, establish similar contact with the surface of various objects. The repetitive contacts with the material lead to the sharpening of our motor imagination.

feeling and inverse imagination | How do we come to perform new, unexpected movements if sensation and movement seem to determine each other? Normally the anticipated sensation of the slippery surface of the handrail or the tile floor induces a definite kind of motor response. If there is a deviation from the expected and actual movement, says Palágyi, the source of the novelty is our feeling. It is so, because sensations awaken in us feelings and these lead to new imagined movement patterns.[8]

In other words, if our feelings undergo some modifications, the change spurs movement variations. The imagined movement is then "deflected," as Palágyi puts it. The process of "deflection" (*Ablenkung*) plays a significant role in our life, especially in our creative activities. Because feelings have such a deflecting power, our body produces not only "direct images" but also "indirect" ones. Direct images correspond to the situation as we experience it. The indirect ones comprise movements that, under the influence of feelings, we recall or invent. Because we find ourselves overcome by feelings, we are able to exert only a partial influence on the shift from direct to indirect images.

For no apparent reason, the bodily contact with an object can give rise to unexpected feelings that lead to surprising movement images. We are inclined to infer the existence of these indirect images when we realize that our body has performed a surprising movement. Circularity exists on this level too: feelings produce unexpected movement images and these arouse in us new feelings. We can now better understand what prompts some musicians to introduce into their play startling deviations. The movements of their fingers trigger momentary feelings and the emotional responses to these movements modify their touch.

Feelings affect our relationship to our past. Inducing indirect images and the subsequent modification of both sensation and movement, they revive past experiences. That is why tactile images lead us not only to the perception of the characteristics of the object, but also to the recovery of past encounters with the same or a different object. Our touch elicits feelings that make us focus on the past rather than on the present.[9] Thus a tactile contact with a piece of stone or wood may lead to the retrieval of diverse textural effects, or it may even prompt the visual memory of the form of the object. When we produce the visual representation of a previously seen tree or a ball, we do this, most often unconsciously, with the help of imagined hand, leg, or eye movements.

Jean Cocteau, in his *Journal d'un inconnu*, described how his tactile movements prompted the representation of some familiar objects. He went back to the street in Paris where he had spent a large part of his childhood. By trailing his fingers along the houses, fences, and walls, he hoped to retrieve some of the pleasant memories of former times. His attempt did not yield a satisfactory result: his hands merely perceived the unevenness of the surface. He suddenly realized that he had to bend down and extend his hand at a lower level, as if he was still a child. A change of bodily posture, allowing him to retrieve different tactile sensations, opened the floodgate to the past: "Just as the needle picks up the melody from the record, I obtained the melody of the past with my hand. I found everything: my cape, the leather of my satchel, the names of my friends and of my teachers, certain expressions I had used, the sound of my grandfather's voice, the smell of his beard, the smell of my sister's dresses and of my mother's gown."[10]

The hand trailing the wall yields not only visual but also nasal and aural memories. I have previously alluded to a direct link between sound and movement: the rhythmic succession of tones compels us to move. The perception of tones elicits in us virtual movements that give rise to actual dancing, marching, or simply humming. Conversely, we can hear sounds from the distant past if, by virtually moving our sound-producing organs, we are able to reproduce them.

The imaginative contact between the liquid and our tongue provokes the memory of the taste of a wine. Overpowered, our sensitive nose seems to be passively delivered to the influence of the abundance of olfactory impressions. Yet, the detection of smell is affected by various factors and, here too, movements could easily facilitate, or hinder, the olfactory process. The keenness of smell develops over time, mainly through an active encounter with a great variety of objects – flowers or wines, for example. Therefore, the motions of sniffing and inhaling that we execute in imagination may prompt the memory of the odour of burning leaves. We are able to recall the smell of an old book if our nose virtually approaches its open pages.

All these imagined movements, as we have just described, cannot be divorced from the upsurge of our feelings. Readers who were delighted by Marcel Proust's great novel would surely consider smell and taste as the ultimate and most resilient access to the past. Yet musical motives and tactile contacts are no less effective in conjuring up the memory of some distant experiences. Because of their close ties to feelings, they provide powerful cues to recall the past. To walk with bare feet on sand or pebbles immediately brings back some of our pleasant childhood memories. Indeed, as I remarked before, touching creates a more intimate contact with objects or people than does vision. The qualities detected by our hands resonate in us and produce a significant and personal echo in us. A friend's delicate touch on our face or elbow heightens our emotions. Buytendijk shrewdly pointed out that we like to describe lasting dispositions through qualities of the touch: "it is not by accident that qualities such as warm, cool, chilly and cold, sharp and obtuse, strong and weak, rough, soft, smooth, slippery and sticky may be used as qualifications for human behavior and so-called characters."[11]

With feelings aroused, tactile contacts perhaps make us more attentive to the values that we detect in some objects. Hence our attraction for old books or antique furniture. Hence also the easy adaptation of the musician's body to a new instrument. The hands and legs of the organ player touch the pedals and manuals and also recognize certain emotional and musical values "produced" by the organ pipes. His play is then considerably shaped by the emotional experience that the tactile contacts elicit, and the music.[12]

creative hands | Our bodily imagination allows us to place ourselves into a new situation. We can transcend the existing present conditions and regulate our behaviour accordingly. "It is an unparalleled wonder that life, without abandoning the place it occupies, can nonetheless behave in such a way as if it had escaped to another place in space and time."[13] If one of the fundamental features of the living body is the ongoing tendency to transcend its physical limit, this openness is in part made possible by the "imaginative flight" from any momentary dwelling place.

Arnold Gehlen considers this ecstatic relationship one of the chief requirements of human survival: we must continually free ourselves from the limits imposed by a given situation in order to satisfy our needs.[14] He also points out that our imagination endows things with "tactile values" – symbols of tactile sensation such as weight, consistency, or temperature. We must be aware of, and expect, tactile sensations in order to execute and refine a vast range of motor skills. It is not only what we touch with our hands or legs that is invested with some value, but also what we are able to reach with the help of some intermediary object or instrument. The billiard player's shot does not merely direct the trajectory of the cue ball, but also anticipates a particular tactile contact that generates the further movement of the object ball. The hand senses the impact in advance and moves the cue accordingly.

Henri Focillon, for his part, asserted that there could be no real knowledge of the world without the "tactile flair" of the hand:

The hand knows that an object has physical bulk, that it is smooth or rough, that it is not soldered to heaven or earth from which it appears to be inseparable. The hand's action defines the cavity of space and the fullness of the objects that occupy it. Surface, volume, density and weight are not optical phenomena. Man first learned about them between his fingers and the hollow of his palm. He does not measure space with his eyes, but with his hands and feet. The sense of touch fills nature with mysterious forces. Without it, nature is like the pleasant landscapes of the magic lantern, slight, flat and chimerical.[15]

But we can attribute to our tactile imagination an inventive function as well. This resides essentially in the human capacity to make use of past encounters with things, experimenting with them, and testing hypotheses.

Music comes from, and calls for, tactile contact with an instrument. The tactile anticipation of hitting a key on the piano is closely connected to the hearing and anticipating of the sounds. The hand anticipates and imagines both sensations and movements to the same extent as the ear hears, in advance, the various characteristics of music, though the hand's projections usually remain unnoticed. I believe that musicians introduce some improvisatory elements into their play – they slightly change the rhythm, intensity, or colour – by allowing their hands to rely on their own formative power. Yet some, much like certain painters, fearing that their art might be considered as a purely manual occupation, like to stress the "vital role of the mind." But music cannot merely be the outcome of a series of premeditated orders. The unconscious anticipation of the hands is just as indispensable to music-making as is the ability to phrase a melody or hear chords. During the learning process of a piece, it cannot be ignored. If a young cello student does not know when and how to use the vibrato, it could be helpful to allow the fingers to select the appropriate rocking motion. "Let the body take over," as Barry Green suggests in his useful book.[16] The colouring of the note could become more natural and distinctive if it is generated by the inventive power of the forearm and the fingers.[17]

Palágyi was well aware of the importance of movement imagination in artistic creations.[18] He carefully observed some Hungarian painters at work in their studios. He realized that no artist could neglect the images "issued" by the hand. Some were able to evoke, on a flat surface, the illusion of plasticity and depth. They created the impression of movement, volume, weight, sensuousness, strength, energy and, above all, life. Those, however, who allowed the inertia of their tactile imagination produced "ethereally tottering, immaterial, and false" works.[19]

With sculpture, the artist's tactile imagination imparts the material with the characteristics of volume, density, palpability, weight, and inner vitality. The stone or iron is then vibrant with desire and intention, acts of its own accord, moves in spite of its immobility. Curiously, some sculptures are filled with liveliness and energy and manifest a peculiar reciprocity: the sculptor's tactile virtuosity carves figures that also seem to possess a sense of touch. In a Romanesque church, when we look at the carved animals biting each other, we realize how the artists were able to endow the stone with one of the fundamental aspects of both human and animal life – the touch. We find the same mysterious liveliness in works that represent the loving solicitude of the hands and the gentleness of the face. Some figures seem to speak to the spectator, or even express a wish or concern. If we have a desire to return to these sculptures, our admiration is, without a doubt, due to our discovery of a breath of life that the artist's hand was able to imprint into the inert matter.[20]

Making a sculpture out of stone essentially consists of creating a form with the help of various tools, above all chisels and hammers. Sculptors carve the material, cut out pieces, and hollow out cavities. By taking away larger and smaller quantities from the block of stone and moving from the exterior to the interior, they progressively create a form. Once again, the anticipation of tactile sensations and the corresponding imagined movements by the hand are the constitutive elements used while making repeated contacts with the stone.

"The hand," insists Focillon, "is not the mind's docile slave."[21] To better understand the hands' contribution, we are advised to close our eyes and caress a carved surface. Now we feel the round and angular aspects of the material, the variation of levels, shapes, and textures. Rudolf Arnheim wrote a descriptive narrative of his hand running over a sculpted head: "Then my hands move along the curves of the jaws converging toward the bow of the chin, which approaches me. The roundness of the chin is paralleled by that of the lips. The nose, finely honed, moves toward me, and the cheeks converge in the shallow depressions I know are the place of the eyes, although these depressions hardly indicate the eyes' presence."[22] Such an exploration reveals those dynamic tensions and particular emphasis that the tactile imagination conveys to the stone during the process of creation. The hand is able to simplify or amplify some motives or, by reproducing certain details, to highlight the individual physiognomy and expression of a figure.

Gert Selle, in his book on the education of the arts, relates a somewhat similar experience.[23] He recommends that students close their eyes and grab and crumble a hunk of moist peat moss. As the material is pressed, or tapped upon, the hands receive a great variety of tactile sensations and movement suggestions. Without opening their eyes, the students are asked to represent on a canvas all the remembered sensations – the resistance, roughness, or softness of the material. Such an exercise helps their sensitive fingers to anticipate tactile qualities and recall them through the practice of drawing forms. Tactile imagination has a dual function: it helps to preserve experiences and to project movements and sensations. "The hands are not simply unskilful and clumsy. They must be considered wise as they commit an act of sabotage against the dictate of the disembodied look. They often resist it, in as much as they can hardly hold the crayon correctly, not to mention the inhibited gestures and the lines that convince the eyes. Thus they make everything incorrectly, as if they knew everything better than the stupid, prejudiced head."[24] Drawing is not merely copying. It is the accomplishment of a "creative mimetic activity," wherein the "embodied tactile imagination" plays a central role. The purpose of blind-drawing is to habituate students to rely on their sense of touch and to produce forms through the fruitful collaboration of their "seeing hands" and "touching eyes."[25]

Prominent theorists of art have long expressed similar views. Focillon remarked that the artist's hand does not merely translate a detailed inner representation. Rather, it displays an uncommon sense of daring and exploring. Endowed with a "magic power," an "unheard-of assurance," it "seems to gambol in utter freedom" and produces the most diverse and astonishing forms: "It searches and experiments for its master's benefit; it has all sorts of adventures; it tries its chance."[26]

There are moments, observes Géza Révész, when we would rather surrender to the initiatives and skills of our hands than to follow the dictates of our intellect. We have the impression that "the hand is more intelligent than the head and is endowed with a greater creative power."[27] Whether we play on the piano or carve a figure out of wood, the hand suggests corrections and innovations. Whatever result we may achieve, concludes Révész, our work arises out of a "cross-fertilizing" interplay of the hand and the mind.[28]

The significance of this relationship was recognized even outside the artistic domain. Alfred North Whitehead explained the weakening of the ability to innovate by the absence of the reciprocal influence of the hand and the brain.[29] The "disuse of hand-craft" is the direct consequence of the embedded tendency to oppose the body to the mind. It is reinforced by the prevalent view that to see is much more important than to touch.

Why does an activity carried out by the hand promote original ideas? Unless we are called upon to perform a repetitive task, manual activities involve presence and require attentiveness. Rarely are we so much ourselves as when absorbed in a captivating task, tinkering with a motor, planting flowers, or playing on a musical instrument. Because of its close ties to feelings, our tactile sense is not merely receptive, but also inventive. The curiosity of children finds its first and perhaps strongest expression in the restless movements of their hand. The desire to palpate, lift, or press objects nurtures in them an attitude of "let's see what happens or how it unfolds," essential for all subsequent artistic and scientific endeavours.

For all those who attempt to kindle the innovative spirit of youth by the acquisition and improvement of computer literacy, I would recommend the writings of painters and sculptors. They present some convincing views about the dynamics of creativity and, by doing so, they make us realize that the "decay of innovation" can be prevented, or reversed, by cultivating what is the closest to us: our hand.

conclusion

In a beautiful passage, Merleau-Ponty compares the body to language. Like the spoken word, the body is neither a pure instrument, ready to obey, nor an end in itself, wanting only to govern. Nevertheless, there are moments when our body, without any purpose, enjoys its autonomy and power. "Sometimes – and then we have the feeling of being ourselves – it lets itself be animated and becomes responsible for a life which is not simply its own. Then it is happy or spontaneous, and we with it."[1] The source of our enjoyment is, above all, the feeling of harmony and unity due to the disappearance of the usual "I and body" tension. All too often, especially during illness, the body is perceived as an instrument to manipulate or an obstacle to overcome. Early in life, we learn how to bridle its impulses, needs, and desires and, because so much emphasis is placed on self-control, we tend to ignore its subtle messages and resonances. If, however, we value and seek awareness, we then experience our body as a source of power and possibilities, a dynamic unfolding of original performances, a creative force that responds judiciously to any perturbation and explores new movement combinations.

Extensive recent research has shown that, at different times, we do have such a pleasant perception of our activity. The terms "peak experience," "flow," or "movement meditation" are used to describe this highly pleasurable state.[2] Billiard players call it "dead stroke." It occurs when we are engrossed in an activity and the accomplishment of the movement itself becomes an intrinsically rewarding experience. In this particular moment, as the will to control the body disappears and the execution of the movement becomes effortless and flawless, we experience the delightful sensation of total involvement and the feeling of harmony and oneness with the different aspects of the motor situation.

An important element of these experiences is what has been called "transcendence of self" or "loss of ego." These expressions do not denote a dreamlike or unconscious state, but rather a non-critical, non-evaluating, fearless, relaxed attitude, a complete absorption in the task at hand. Without calculating the possible result, we execute the appropriate movements. We have the impression of being carried by our vital energies and moving on autopilot, so to speak. Enjoyment, in this sense, is not synonymous with health or fitness, but springs from a motor experience that leads to a heightened awareness of our autonomous body. It may occur while we sing, play enthralling tennis matches, or walk along a beach.

Indeed, we come to experience pleasant sensations during the most diverse activities. One of them is lovemaking. The delight that one feels does not depend on some physiological processes alone, but rather on how one experiences one's body and relates to another. As Gernot Böhme has remarked, a sexual encounter elicits an unusual experience: the body reaches an altogether "other state" (*andere Zustand*).[3]

In our daily life, in a state of well-being, we identify ourselves with whatever lies beyond and ahead of us. Whether concerned with our projects, tasks, or goals, we may recognize that our body supports our intentions, yet we perceive ourselves as if we were "beyond it." We view it as a silent medium that allows us to reach our objectives and meet the demands of our everyday existence. If, on the other hand, we become tired, exhausted, ill, or are unable to correctly perform a task, we become conscious of our body. It appears in its materiality as an obstacle, making impossible the fulfilment of our plans. In the state of "ill-being," the body manifests its unavailability. It is no longer the imperceptible support that we vaguely sense, but an object asserting itself with a quasi-independence. It resists our efforts and desires. It is not what we are, but what we have.[4] In the "other state," we do become aware of our body, but we don't experience it through our objectifying consciousness. As our consciousness submerges into the body, we live and feel it as if it were from inside. Such an inward awakeness may be achieved in lovemaking, dance, sport, meditation, or even breathing. While breathing, we may become aware of the inner forces working inside us and develop an "inward consciousness" of our body.[5] Eugen Herrigel, in his famous work, describes how archery cultivates the "exquisite state of unconcerned immersion in oneself."[6] Here too, through the concentration on breathing movements, we come to a "primordial state." We feel all our energies from inside and are able to mobilize them at any moment with "rapturous certainty."[7]

Gernot Böhme writes about the "extraordinary significance" of bodily love.[8] It allows us to achieve what usually eludes us under our contemporary living conditions, namely the experience of wakeful relaxation of, and enjoyable unity with, the body. We are always able to consider the acts of walking or drinking as if they were carried out by someone else. This is no longer possible in lovemaking. Here, as Jan Kott points out, the alienation of the self from the body is impossible. "It is then that the soma and the anima are one. When you can no longer extricate yourself from yourself, the experience is no longer someone's else."[9]

There are other occasions when we are truly pressed to give way to the body's ability to perform the appropriate behaviour. This occurs when we burst into tears.[10] When we cry, we neither confront and use our body as an instrument nor express our inner state in an articulated manner. The self-control and self-transparency are no longer available means to deal with the situation. We find ourselves in an "other state," since, in the presence of a "constraining power," our body provides "an autonomous reaction."[11] The cardinal element of this experience is the act of surrender: letting ourselves be overcome by, or dissolved in, weeping; we yield to our body and allow its resources to respond.

In dance or play, our movements are an integral part of a more encompassing experience of space and time. The characteristics of the lived space of dance, defined by the absence of goals, specific directions, and fixed axes, promote and facilitate the experience of unity with the body and the surrender to its capabilities.[12] The disappearance of purposive attitude changes the spatial structure to the same extent that dissolution of all oppositions heightens the state of alert but relaxed receptiveness. Without a concern for specific goals, we find ourselves in synchrony with the surroundings, consisting, above all, of the experience of equivalence and reciprocity.

As we establish a more intimate contact with space, we also come to free ourselves from the usual way of experiencing time. Our present is no longer an incomplete sequence that receives its significance only from future and past events. It is rich and complete in itself, almost as if it has some sort of extension or density. Stepping outside the objective and general order of time, we experience the present as a deeply satisfying moment of our personal becoming. However, neither the past nor the future loses its importance. The past is eminently there through the conservation of our acquired skills and incorporated values. This memory is also prospective since it allows us to improvise and invent new solutions. The future offers us not merely fixed tasks and limitations but, above all, possibilities to create and progress.

The value of the exquisite state has already been recognized, about two hundred years ago, by Heinrich von Kleist. In a remarkable short essay, he warns his readers about the undesirable consequences of excessive reasoning and calculation, and invites them to reflect after an action, and not before. Reflection impedes the harmonious unfolding of movements and hinders the beneficent working of bodily energies. Just like the wrestler who, in order to win, must act according to the "prompting of the moment," likewise, we should respond to unexpected challenges by renouncing hard and lengthy thinking and confidently relying on a power that springs from our bodily feelings. "Life itself is a struggle with Fate; and in our actions it is much as it is in a wrestling match.... A man must, like that wrestler, take hold of life and feel and sense with a thousand limbs how his opponent twists and turns, resists him, comes at him, evades him and reacts: or he will never get his way in a conversation, much less in a battle."[13] Kleist succinctly presents a paradox that, if properly applied, brings success to many human endeavours. The more our action is guided by our conscious will, the less we succeed. When we desperately try to ski or dance well, we make one mistake after another. It is, therefore, advisable to proceed in a reverse order. At the moment of action, we should abandon all attempts to consciously control our movements and trust the various resources of our body. In other words, we should get rid of all tense effort and purposeful planning, and not interfere with the infallible functioning of our body – the "splendid feeling" (*herrliche Gefühl*) inhabiting our clever body. All careful analysis of the implemented solutions should follow our action. This will serve to improve the spontaneous functioning of our body and prepare us to face new challenges.

In a delightful short story about play, Walter Benjamin presents a different view on our unconscious knowledge that is converted into movements. Playing in the casino is much like facing a danger: the body deals with the situation and disregards the dictates of the mind. (These kinds of challenges require bodily presence rather than presence of mind.) While placing a bet, the hand successfully guesses the winning number if the consciousness does not intervene. As soon as the player starts to "think right," the hand inevitably makes mistakes. "Play is a disreputable occupation because what makes the organism perform the most subtle and precise actions also provokes conscienceless behaviour."[14]

In another context, however, the value of careful thinking is assessed differently. William James believed that the chief problem with teaching, as well as with many other social activities, is that we tend to be too preoccupied with the results. We are too cautious about making a mistake and, consequently, calculate what we say or do. Hence the importance of forgetting all our worries, "taking the brakes off the heart" and trusting our spontaneity. The remedy for the tension and excessive self-consciousness lies in the recovery of a life-style that considers ease and relaxation as central values.[15]

How can we promote such a life-style? How can we achieve a more intimate and trustful contact with our surrounding world and foster some of our bodily abilities?

A life according to the gospel of surrender can hardly be achieved at will. Adults and youth need adequate opportunities to experiment with movements and take initiatives freely, without fear and constraint. The creation of a free space – a leisurely space for innovation – is what an educational system has to promote. Children could then learn early enough to combine concentration with relaxation, a paradoxical art that is, according to Aldous Huxley, the key to reaching proficiency in any field.[16] It is through concentration that they immerse themselves in an activity and fend off all external disturbances. And it is through "wise passiveness" that they remove the barriers of calculating consciousness and allow the unrestricted functioning of their bodily powers.

Philosophers, educators, and even scientists plead, alas repeatedly in vain, in favour of an education where various bodily activities play a central role. They rightly claim that productive thinking cannot be confined to one or two specific subjects in the curriculum. The creative process grows out of the whole living organism. The practice of the arts could heighten the pupils' sensitivity to the new and original, and make possible the use of self-expressive capabilities far beyond the realm of the school.

The introduction of singing or dancing into the educational systems is not enough. An altogether different understanding of the body is required. We should stop considering the body as an instrument, a machine, or an object of possession that responds to, or resists, external challenges. In many instances, our body is a partner; our creative potentials and already acquired versatile technical knowledge offer their support for our endeavours.[17] There is a specific freedom of the body that becomes manifest in the astonishing independence of the hands and legs. They move not only with a remarkable agility, but also with a surprising ease. Hans-Eduard Hengstenberg sees in the body's ease a genuine virtue that, in our abstract and standardized societies, we tend to lose.[18] This virtue consists of the remarkable ability of the body to mould its various parts into different shapes, move according to a variety of rhythms, and adopt new types of symbolic behaviour. It allows us to sense the prevailing requirement of a situation and form our body accordingly.

We then adopt a particular attitude that we might characterize with the expression of *syntony*. Syntony refers to the fundamental faculty of establishing a harmonious and sympathetic contact with the everyday world. We attain an accord between ourselves and whatever we are dealing with – the tree we prune, the origami crane we make, or the food we arrange on a plate. Syntony pertains, beyond the experience of vibrating in unison with something, to the character of the contact we establish with our own body. We no longer confront it, but allow its impulses, energies, and clever capabilities to guide our action.

Some individuals are very much in tune with their environment.[19] They live with sympathetic ties to their surroundings because they acquired early enough the art of "getting out of the way," the consideration of their carrying body as a partner, and the ability to yield to its autonomous and creative functioning. To my mind, imparting this art to our children is one of the chief tasks of all education.

notes

introduction

1 Max Horkheimer and Theodor W. Adorno, *Dialectic of Enlightenment: Philosophical Fragments*, trans. John Cumming (New York: Continuum, 1989), 232.
2 Arnold Gehlen, "Neuartige kulturelle Erscheinungen," in *Anthropologische und sozialpsychologische Untersuchungen* (Reinbek: Rowohlt Verlag, 1993), 163–67.
3 Albert Borgmann, *Crossing the Postmodern Divide* (Chicago: University of Chicago Press, 1992), 83.
4 Albert Borgmann, *Technology and the Character of Contemporary Life. A Philosophical Inquiry* (Chicago: University of Chicago Press, 1984), 114–24.
5 Shoshana Zuboff, *In the Age of the Smart Machine: The Future of Work and Power* (New York: Basic Books, 1988), 75.

6 Daniel J. Boorstin, *The Image or What Happened to the American Dream* (Harmondsworth: Penguin, 1961), 86–125.
7 Albert Borgmann, *Technology*, 56.
8 Richard Sennett, *Flesh and Stone: The Body and the City in Western Civilization* (New York: W. W. Norton, 1994), 18.
9 Robert J. Yudell, "Body Movement," in Kent C. Bloomer and Charles W. Moore, *Body, Memory, and Architecture* (New Haven: Yale University Press, 1977), 72.
10 Borgmann, *Crossing the Postmodern Divide*, 106.
11 See Christopher Lasch, *The Culture of Narcissism: American Life in an Age of Diminishing Expectations* (New York: W. W. Norton, 1991), 184.
12 R. D. Laing, *The Divided Self: An Existential Study in Sanity and Madness* (Harmondsworth: Penguin, 1965), 67, 69.
13 Cornelius A. van Peursen, *Body, Soul, Spirit: A Survey of the Body-Mind Problem*, trans. Hubert H. Hoskins (London: Oxford University Press, 1966); Medard Boss, *Existential Foundations of Medicine and Psychology*, trans. Stephen Conway and Anne Cleaves (New York: Jason Aronson, 1984), 100–105. See also Henk Ten Have, "The Anthropological Tradition in the Philosophy of Medicine," *Theoretical Medicine*, 16 (1994): 3–14; Richard M. Zaner, "The Discipline of the 'Norm:' A Critical Appreciation of Erwin Straus," *Human Studies*, 27 (2004): 37–50.
14 Maurice Merleau-Ponty, *Phenomenology of Perception*, trans. Colin Smith (London: Routledge, 1994), 207.
15 See, on this topic, the classic work of Walter B. Cannon, *The Wisdom of the Body*, 2nd ed. (Magnolia, MA: Peter Smith, 1978); and the more recent monograph of Sherwin B. Nuland, *The Wisdom of the Body* (New York: Alfred A. Knopf, 1997).
16 Jan Hendrik van den Berg, *A Different Existence: Principles of Phenomenological Psychopathology* (Pittsburgh: Duquesne University Press, 1995), 51.
17 Frederik J. J. Buytendijk, *Prolegomena einer anthropologischen Physiologie* (Salzburg: Otto Müller Verlag, 1967), 62–63.
18 Etienne Gilson, *Painting and Reality* (New York: Pantheon Books, 1957), 31.
19 Ibid., 35.
20 Buytendijk, *Prolegomena*, 227.
21 F.J.J. Buytendijk, "Das Menschliche der menschlichen Bewegung," in *Das Menschliche: Wege zu seinem Verständnis* (Stuttgart: Koehler Verlag, 1958), 184.
22 Aldous Huxley, "The Education of an Amphibian," in *Adonis and the Alphabet and Other Essays* (London: Chatto & Windus, 1956), 9–38; "Education on the Nonverbal Level," *Deadalus* 91 (1962): 279–93; "The Ego," in *The Human Situation: Lectures at Santa Barbara, 1959*, edited by

Piero Ferrucci (London: Flamingo Modern Classic, 1994), 129–42. See also Huxley's foreword to Luigi Bonpensiere, *New Pathways to Piano Technique: A Study of the Relations between Mind and Body with Special Reference to Piano Playing* (New York: Philosophical Library, 1967), v–xiii.

23 Huxley, "The Education of an Amphibian," 26.

24 Frederik J. J. Buytendijk, Erwin W. Straus, Eugène Minkowski, Viktor Emil Freiherr von Gebsattel, Jürg Zutt, Paul Christian, Herbert Plügge, Hubertus Tellenbach, and Jan Hendrik van den Berg are among the most important figures of this movement.

25 The archaeologist Bjørnar Olsen, in his recent study, alleges that most of the scholars of social sciences and humanities show no interest for the materiality of our everyday life. He fails to notice, however, that the anthropologically oriented thinkers have always emphasized the inseparability of bodily actions from the material context and refused to treat the experiencing men and women as "extramundane subjects" (Straus). See "Material Culture after Text: Re-Membering Things," *Norwegian Archaeological Review* 36, no. 2 (2003): 87–104.

26 See van den Berg, *A Different Existence*, 4.

chapter 1

1 See Gerd Haeffner, *Philosophische Anthropologie*, 3rd rev. ed., vol. 1 of *Grundkurs Philosophie* (Stuttgart: Verlag W. Kohlhammer, 2000), 136–38.

2 See Erwin W. Straus, *The Primary World of Senses: A Vindication of Sensory Experience*, trans. Jacob Needleman (New York: Free Press of Glencoe, 1963), 194–96.

3 See F.J.J. Buytendijk, *Wesen und Sinn des Spiels: Das Spielen des Menschen und der Tiere als Erscheinungsform der Lebenstriebe* (1933), reprint (New York: Arno Press, 1976), 62–79.

4 F.J.J. Buytendijk, *Allgemeine Theorie der menschlichen Haltung und Bewegung* (Berlin: Springer-Verlag, 1956), 296–98.

5 Eugenio Barba, "The Dilated Body," in *A Dictionary of Theatre Anthropology: The Secret Art of the Performer*, eds., Eugenio Barba and Nicola Savarese (London: Routledge, 1995), 54.

6 Michael Chekhov, *To the Actor* (London: Routledge, 2002), 3.

7 See Philipp Lersch, *Aufbau der Person*, 11th ed. (Munich: Johann Ambrosius Barth, 1970), 190–93.

8 Gregory Bateson, "Conscious Purpose versus Nature," in *Steps to an Ecology of Mind: Collected Essays in Anthropology, Psychiatry, Evolution, and Epistemology* (Northvale, NJ: Jason Aronson, 1987), 444.

9 See Haeffner, *Philosophische Anthropologie*, 138.

10 Paul Ricoeur, *Freedom and Nature: The Voluntary and the Involuntary*, trans. Erazim V. Kohák (Evanston: Northwestern University Press, 1966), 275–76.

11 See Ernst Bloch, *The Principle of Hope*, vol. 1, trans. Neville Place, Stephen Place and Paul Knight (Cambridge, MA: MIT Press, 1986), 47–50.

12 Jürg Zutt, "Über den tragenden Leib," in *Auf dem Wege zu einer anthropologischen Psychiatrie: Gesammelte Aufsätze* (Berlin: Springer-Verlag, 1963), 419–20.

13 See Felix Hammer, *Leib und Geschlecht: Philosophische Perspektiven von Nietzsche bis Merleau-Ponty und phänomenologisch-systematischer Aufriss* (Bonn: Bouvier Verlag, 1974), 192–96.

14 Buytendijk, *Prolegomena*, 134. Werner Herzog, in his film *Little Dieter Needs to Fly* (1997), shows how meaning is bestowed upon specific objects (doors) on the basis of one's life experience.

15 See Herbert Plügge, *Wohlbefinden und Missbefinden: Beiträge zu einer medizinischen Anthropologie* (Tübingen: Max Niemeyer Verlag, 1962), 91–106.

16 Hubertus Tellenbach, *Melancholy: History of the Problem, Endogeneity, Typology, Pathogenesis, Clinical Consideration*, trans. Erling Eng (Pittsburgh: Duquesne University Press, 1980), 17–57; "Die Begründung psychiatrischer Erfahrung und psychiatrischer Methoden in philosophischen Konzeptionen vom Wesen des Menschen," in *Philosophische Anthropologie: Erster Teil*, eds. Hans-Georg Gadamer and Paul Vogler (Stuttgart: Georg Thieme Verlag; Munich: Deutscher Taschenbuch Verlag, 1974), 169–75.

17 Buytendijk, *Prolegomena*, 43.

18 See Tellenbach, *Melancholy*, 41.

19 Tellenbach, "Die Begründung," 170.

20 Tellenbach, *Melancholy*, 42.

21 Ibid., 41.

22 Ibid., 42.

23 Buytendijk, *Allgemeine Theorie*, 288.

24 Claude Bruaire, *Philosophie du corps* (Paris: Seuil, 1968), 148–51.

25 John Blacking, "Towards an Anthropology of the Body," in *The Anthropology of the Body*, ed. John Blacking (London: Academic Press, 1977), 11.

chapter 2

1 Maurice Merleau-Ponty, *The Structure of Behavior*, trans. Alden L. Fisher (Boston: Beacon Press, 1967), 173.

2 See Henri Maldiney, "Comprendre," *Revue de Métaphysique et de Morale* 1–2 (1961): 52–53; Michael Polanyi, *The Tacit Dimension* (New York: Anchor Books, 1967), 1–25.
3 See Henri Maldiney, *Penser l'homme et la folie: À la lumière de l'analyse existentielle et de l'analyse du destin* (Grenoble: Jérôme Millon, 1991), 295–323.
4 Erwin W. Straus, *Man, Time, and World: Two Contributions to Anthropological Psychology*, trans. Donald Moss (Pittsburgh: Duquesne University Press, 1982), 60–61.
5 Erwin W. Straus, "The Forms of Spatiality," in *Phenomenological Psychology* (1966), reprint, trans. Erling Eng (New York: Garland, 1980), 3–37; *The Primary World of Senses*, 367–79. See also Renaud Barbaras, "Affectivity and Movement: The Sense of Sensing in Erwin Straus," *Phenomenology and the Cognitive Sciences* 3 (2004): 215–28.
6 Straus, "Forms of Spatiality," 27.
7 Buytendijk, *Prolegomena*, 51; Jean Ladrière, "La ville, inducteur existentiel," in *Vie sociale et destinée* (Gembloux: Duculot, 1973), 139–60.
8 Michel Henry, *The Essence of Manifestation*, trans. Girard Etzkorn (The Hague: Martinus Nijhoff, 1973), 459–68.
9 Ladrière, "La ville, inducteur existentiel," 154.
10 Louis Lavelle, *The Dilemma of Narcissus*, trans. W. T. Gairdner (London: Allen & Unwin, 1973), 84.
11 See Eugène Minkowski, "Voyons-nous avec les yeux ?" in *Vers une cosmologie: Fragments philosophiques* (Paris: Aubier-Montaigne, 1967), 133–34.
12 C. S. Lewis, *Studies in Words*, 2nd ed. (Cambridge: Cambridge University Press, 1967), 159.
13 Lavelle, *Dilemma of Narcissus*, 84.
14 Paul Weiss also defines sensibility as a singular capacity to discriminate. It differs from sensitivity inasmuch as it is more refined and relies on the proper functioning of different parts of the body. "Living bodies are responsive. Their responsiveness is the product of an exercise of their of their existence, differently answering to the different pressures and occurrences encountered. Their existence is a sensitive power. When the body is merely alive, this power serves to sensitize it, have it ready to respond. When the sensitive power is expressed to a greater degree than this it makes the body sensible, a being which differentially acts as a single body with stresses of various sorts in different parts of that body, to make possible a more flexible and appropriate response." "Man's Existence," *International Philosophical Quarterly* 1 (1961), 561. See also his *Privacy* (Carbondale: Southern Illinois University Press, 1983), 57–73.
15 Straus, *The Primary World of Senses*, 213–14.

16 F.J.J. Buytendijk, *Le football: Une étude psychologique* (Paris: Desclée de Brouwer, 1952), 25.

17 Jean Nogué, *Esquisse d'un système des qualités sensibles* (Paris: PUF, 1943), 109; see also F.J.J. Buytendijk, "Some Aspects of Touch," *Journal of Phenomenological Psychology* 1 (1970): 114.

18 Straus, *The Primary World of Senses*, 254.

19 Ibid., 367–79.

20 Anthony Storr, *Music and the Mind* (New York: Free Press, 1992), 26.

21 See, for example, Robert Rivlin and Karen Gravelle, *Deciphering the Senses: The Expanding World of Human Perception* (New York: Simon and Schuster, 1984).

22 David Katz, "The Vibratory Sense," in *The Vibratory Sense and Other Lectures* (Orono, ME: University Press, 1930), 90–103. See also Jacqueline Verdeau-Paillès, "Music and the Body," in *The Fourth International Symposium on Music in Rehabilitation and Human Well-Being*, ed. Rosalie Rebollo Pratt (Lanham, MD: University Press of America, 1987), 42–43.

23 Nogué, *Esquisse d'un système*, 180.

24 René A. Spitz, *The First Year of Life: A Psychoanalytic Study of Normal and Deviant Development of Object Relations* (New York: International Universities Press, 1965), 134.

25 Daniel N. Stern, *The Interpersonal World of the Infant: A View from Psychoanalysis and Developmental Psychology* (New York: Basic Books, 1985), 67.

26 Spitz, *The First Year of Life*, 136.

27 See Anthony Storr, *The Dynamics of Creation* (London: Secker & Warburg, 1972), 189–91.

28 Alfred Schutz, "Making Music Together. A Study of Social Relationship," in *Studies in Social Theory*, vol. 2 of *Collected Papers*, ed. Maurice Natanson (The Hague: Martinus Nijhoff, 1964), 159–78.

29 Storr, *Music and the Mind*, 124.

30 See Maurice Merleau-Ponty, "Man and Adversity," in *Signs*, trans. Richard C. McClearly (Evanston: Northwestern University Press, 1964), 234–35.

31 Merleau-Ponty, *Phenomenology of Perception*, 327.

32 Ibid., 281.

33 Tony Hiss, *The Experience of Place* (New York: Vintage Books, 1991), 15.

34 Kent C. Bloomer and Charles W. Moore, *Body, Memory, and Architecture* (New Haven: Yale University Press, 1977), 37–56.

35 Ladrière, "La ville, inducteur existentiel," 158–60.

36 Gernot Böhme, *Anthropologie in pragmatischer Hinsicht: Darmstädter Vorlesungen* (Frankfurt am Main: Suhrkamp Verlag, 1985), 199. See

also by the same author "Atmosphäre als Grundbegriff einer neuen Ästhetik," in *Atmosphäre: Essays zur neuen Ästhetik* (Frankfurt am Main: Suhrkamp Verlag, 1995), 21–48.

37 Hubertus Tellenbach, *Geschmack und Atmosphäre: Medien menschlichen Elementarkontaktes* (Salzburg: Otto Müller Verlag, 1968).

38 Tellenbach, "Die Begründung," 175–79.

39 See Eugène Minkowski, "Se répandre (L'olfactif)," in *Vers une cosmologie: Fragments philosophiques* (Paris: Aubier-Montaigne, 1967), 116.

40 Tellenbach, *Geschmack und Atmosphäre*, 47.

41 Nicolaï Hartmann views the acts of "seeing through" (*Hindurchsehen*) and "hearing through" (*Hindurchhören*) and their correlates, the "affective tones" (*Gefühlstöne*), as constitutive elements of human perception. *Ästhetik*, 2nd ed. (Berlin: Walter de Gruyter, 1966), 42–49.

42 A recent study shows that, through the voice attractiveness, men and women detect important information about the body configuration and sexual behaviour of the opposite sex. See Susan M. Hughes, Franco Dispenza and Gordon G. Gallup Jr., "Ratings of Voice Attractiveness Predict Sexual Behavior and Body Configuration," *Evolution and Human Behavior* 25 (2004): 295–304.

43 Minkowski, "Se répandre," 119.

44 Alan Walker, *Reflections on Liszt* (Ithaca, NY: Cornell University Press, 2005), 58.

45 J. Rudert, "Die persönliche Atmosphäre," *Archiv für die gesamte Psychologie* 116 (1964): 295. See also Otto Friedrich Bollnow, *Die pädagogische Atmosphäre: Untersuchungen über die gefühlsmäßigen zwischenmenschlichen Voraussetzungen der Erziehung* (Essen: Verlag Die Blaue Eule, 2001).

46 Stephen Vizinczey, *An Innocent Millionaire* (Boston: The Atlantic Monthly Press, 1985), 321.

47 F.J.J. Buytendijk, "L'objectivité des choses et l'expressivité des formes," *Psychiatria, Neurologia, Neurochirurgica* 73 (1970): 427–31; See also Heinz Werner, *Comparative Psychology of Mental Development*, rev. ed. (New York: International Universities Press, 1980), 389–402.

48 See Rudolf Arnheim, "Art Among the Objects," in *To the Rescue of Art: Twenty-Six Essays* (Berkeley: University of California Press, 1992), 9.

49 See Hans-Georg Gadamer, "The Nature of Things and the Language of Things," in *Philosophical Hermeneutics*, trans. and ed. David E. Linge (Berkeley: University of California Press, 1977), 71.

50 Maldiney, "Comprendre," 75–76.

51 Minkowski, "Se répandre," 118.

52 Hiss, *The Experience of Place*, 34.

53 Paul Valéry, *Cahiers*, vol. 1 (Paris: Gallimard, Pléiade, 1973), 1206.

54 Nils Lennart Wallin, *Biomusicology: Neurophysiological, Neuropsychological, and Evolutionary Perspectives on the Origins and Purposes of Music* (Stuyvesant, NY: Pendragon Press, 1991), 21.
55 Storr, *Music and the Mind*, 184.
56 John Blacking, "The Biology of Music-Making," in *Ethnomusicology*, vol. 1 of *The New Grove Dictionary of Music and Musicians*, ed. Helen Myers (London: Macmillan Press, 1992), 304.
57 Ibid., 306.

chapter 3

1 On the relationship between voluntary disposition and natural spontaneity, see Bruaire, *Philosophie du corps*, 141–64.
2 Arnold Gehlen, *Der Mensch: Seine Natur und seine Stellung in der Welt*, 12th ed. (Wiesbaden: Akademische Verlagsgesellschaft Athenaion, 1978), 132.
3 Paul Christian, "Vom Wertbewußtsein im Tun. Ein Beitrag zur Psychophysik der Willkürbewegung," in Frederik J. J. Buytendijk, Paul Christian and Herbert Plügge, *Über die menschliche Bewegung als Einheit von Natur und Geist* (Schorndorf: Verlag Karl Hofmann, 1963), 19–44.
4 Buytendijk, *Prolegomena*, 188.
5 See Arnulf Rüssel, "Gestalt und Bewegung: Psychologische Grundfragen der Sprechspur," *Psychologische Beiträge* 2 (1955): 425–29; Paul Christian, "Möglichkeiten und Grenzen einer naturwissenschaftlichen Betrachtung der menschlichen Bewegung," *Jahrbuch für Psychologie und Psychopathologie* 4 (1956): 353–54.
6 See Ricoeur, *Freedom and Nature*, 285.
7 Viktor Emil Freiherr von Gebsattel, "Süchtiges Verhalten im Gebiet sexueller Verirrungen," in *Prolegomena einer medizinischen Anthropologie: Ausgewählte Aufsätze* (Berlin: Springer-Verlag, 1954), 187–94. See also van den Berg, *A Different Existence*, 57–58.
8 Straus, "The Forms of Spatiality," 26.
9 Martin Seel, "Die Zelebration des Unvermögen: Zur Ästhetik des Sports," *Deutsche Zeitschrift für europäisches Denken* 47 (1993): 91–100.
10 Ibid., 97.
11 Ibid.
12 Ricoeur, *Freedom and Nature*, 290.
13 F.J.J. Buytendijk, *Mensch und Tier: Ein Beitrag zur vergleichenden Psychologie* (Reinbek: Rowohlt Verlag, 1958), 47.
14 Straus, *The Primary World of Senses*, 363.
15 Ibid., 364.
16 Buytendijk, *Prolegomena*, 193.

17 Eugène Minkowski, "Spontaneity (...Spontaneous Movement Like This!)," in *Readings in Existential Phenomenology*, eds. Nathaniel Lawrence and Daniel O'Connor (Englewood Cliffs, NJ: Prentice-Hall, 1967), 168–77.
18 Ibid., 176.
19 Buytendijk, *Allgemeine Theorie*, 296–98.
20 Minkowski, "Spontaneity," 174.
21 See Peter Röthig, "Betrachtungen zur Körper- und Bewegungsästhetik," in *Grundlagen und Perspektiven ästhetischer und rhythmischer Bewegungserziehung*, eds. Eva Bannmüller and Peter Röthig (Stuttgart: Ernst Klett Verlag, 1990), 94–95.
22 See Buytendijk, *Allgemeine Theorie*, 362.
23 Bruno Nettl, "Thoughts on Improvisation: A Comparative Approach," *The Musical Quarterly* 60 (1974): 12.
24 Jeff Pressing, "Cognitive Processes in Improvisation," in *Cognitive Processes in the Perception of Art*, eds. Ray W. Crozier and Antony J. Chapman (Amsterdam: Elsevier, 1984), 346.
25 Ibid., 353.
26 Yehudi Menuhin, "Improvisation and Interpretation," in *Theme and Variations* (New York: Stein and Day, 1972), 37.
27 Helmuth Plessner, "Zur Anthropologie der Musik," in *Ausdruck und menschliche Natur*, vol. 7 of *Gesammelte Schriften* (Frankfurt am Main: Suhrkamp Verlag, 1982), 198.
28 Heinz Heckhausen, "Entwurf einer Psychologie des Spielens," in *Das Kinderspiel*, ed. Andreas Flitner (Munich: Piper, 1978), 153. See also Ed Sarath, "A New Look at Improvisation," *Journal of Music Theory* 40 (1996): 28–31; Jeff Pressing, "Improvisation: Methods and Models," in *Generative Processes in Music: The Psychology of Performance, Improvisation, and Composition*, ed. John A. Sloboda (Oxford: Clarendon Press, 1988), 139.
29 Barry Green, with W. Timothy Gallwey, *The Inner Game of Music* (New York: Doubleday, 1986), 207.
30 Pressing, "Cognitive Processes in Improvisation," 359.
31 Alfred Pike, "A Phenomenology of Jazz," *Journal of Jazz Studies* 2 (1974): 91.
32 Ibid.
33 Janice E. Kleeman contends that "composers, performers, and audiences all possess an instinctive desire for the new, the unknown, the challenging, which may spring from the evolutionary process: those human beings survive and procreate who best cope with the unexpected in a dangerous world. Whatever its origin, our creative compulsion is a wide vein in the bedrock of habitual behavior." "The

Parameters of Musical Transmission," *Journal of Musicology* 4 (1985/86): 21.

34 David Sudnow, *Ways of the Hand: The Organization of Improvised Conduct* (Cambridge, MA: MIT Press, 1993), xiv.

35 Ibid., 115.

36 Eric F. Clarke, "Generative Principles in Music Performance," in *Generative Processes in Music: The Psychology of Performance, Improvisation, and Composition*, ed. John A. Sloboda (Oxford: Clarendon Press, 1988), 7.

37 See Sarath, "A New Look at Improvisation," 19–23.

38 Pressing, "Improvisation: Methods and Models," 149.

39 Rüssel, "Gestalt und Bewegung," 426–27. On the problem of mental representation, see Anne Jaap Jacobson, ed., *Hubert Dreyfus and Problem of Representation*. Special Issue of *Phenomenology and the Cognitive Sciences* 1, no. 4 (2002): 357–425.

40 Ricoeur, *Freedom and Nature*, 290.

41 A rigid insistence on internal representation could also hinder our understanding of spontaneity. Oliver Sacks strongly criticizes all those who undertake an empirical analysis of spontaneous events and give a mechanical interpretation of "the musicality of action and life." "They speak of 'programs,' 'procedures,' 'solving the motor task' – as if their patients were computers, or 'cyborgs.' They miss the essential beauty and mystery of action, they miss its grace, its musicality." *A Leg to Stand On* (New York: Harper Perennial, 1990), 216.

42 Robert Spaemann, *Basic Moral Concepts*, trans. T. J. Armstrong (London: Routledge, 1989), 76.

43 See van Peursen, *Body, Soul, Spirit*, 181.

44 Eliot Deutsch, *Personhood, Creativity and Freedom* (Honolulu: University of Hawaii Press, 1982), 134.

45 See Charles Taylor, *The Ethics of Authenticity* (Cambridge, MA: Harvard University Press, 1992).

46 See David Heyd, "Tact: Sense, Sensitivity, and Virtue," in *Inquiry* 38 (1995): 217–31; and my "Du tact," in *Science et Esprit* 47 (1995): 329–33.

47 Helmuth Plessner, *The Limits of Community*, trans. Andrew Wallace (Amherst, NY: Humanity Books, 1999), 168.

48 For a remarkable analysis of bodily wisdom, see Thomas De Koninck, *Philosophie de l'éducation: Essai sur le devenir humain* (Paris: PUF, 2004), 36–50. For a specific discussion applied to ethical action, see Eugène Minkowski, in his *Lived Time: Phenomenological and Psychopathological Studies*, trans. Nancy Metzel (Evanston: Northwestern University Press, 1970), 111–21.

chapter 4

1 Helmuth Plessner, "Der imitatorische Akt," in *Ausdruck und menschliche Natur*, vol. 7 of *Gesammelte Schriften* (Frankfurt am Main: Suhrkamp Verlag, 1982), 452.

2 One might question Jean Piaget's assertion that "imitation is never a behavior which is an end in itself." *Play, Dreams and Imitation in Childhood*, trans. C. Gattegno, F. M. Hodgson (New York: W. W. Norton, 1962), 73.

3 F.J.J. Buytendijk, *Traité de psychologie animale*, trans. A. Frank-Duquesne (Paris: PUF, 1952), 335. See also Howard Gardner, *Frames of Mind: The Theory of Multiple Intelligences* (New York: Basic Books, 1985), 226–33.

4 Max Horkheimer, *Eclipse of Reason* (New York: Oxford University Press, 1947), 114.

5 See Eugenio Barba, *The Paper Canoe: A Guide to Theatre Anthropology*, trans. Richard Fowler (London: Routledge, 1995), 25–30.

6 Walter Benjamin, "On the Mimetic Faculty," in *Reflections*, trans. Edmund Jephcott (New York: Schocken Books, 1986), 333.

7 See Stephen Buckland, "Ritual, Bodies and 'Cultural Memory'," in *Liturgy and the Body*, eds. Louis-Marie Chauvet and Francois Kabasele Lumbala (London: SCM Press, 1995), 49–56.

8 Ernst Cassirer, *The Philosophy of Symbolic Forms*, vol. 1, *Language*, trans. Ralph Manheim (New Haven: Yale University Press, 1955), 183.

9 Not only Jean Piaget's detailed study on the syncretism of perception and imitation but also more recent findings confirm the accuracy of Cassirer's observations. Andrew N. Meltzoff and M. Keith Moore showed that children not only select aspects of the perceived gesture but also combine these aspects creatively and end up "constructing" a "novel act." "Infant's Understanding of People and Things: From Body Imitation to Folk Psychology," in *The Body and the Self*, eds. Jose Luis Bermudez, Anthony Marcel and Naomi Eilan (Cambridge, MA: MIT Press, 1995), 52.

10 Margaret Mead, "Balinese Character," in Gregory Bateson and Margaret Mead, *Balinese Character: A Photographic Analysis* (New York: New York Academy of Sciences, 1942), 18.

11 Ibid.

12 David Abercrombie, "Conversation and Spoken Prose," in *Studies in Phonetic and Linguistics* (London: Oxford University Press, 1971), 9.

13 Ludwig Wittgenstein, *Zettel*, trans. G. E. M. Anscombe (Berkeley: University of California Press, 1970), 41e.

14 Abercrombie, "Conversation and Spoken Prose," 6.
15 See Géza Révész, "Die Sprachfunktion der Hand," *Psychologische Beiträge* 2 (1955): 254–65.
16 José Ortega y Gasset, *Man and People,* trans. Willard R. Trask (New York: W. W. Norton, 1963), 92.
17 Jeremy Campbell, "The Conversational Waltz," in *Winston Churchill's Afternoon Nap* (New York: Simon and Schuster, 1986), 237.
18 Harvey B. Sarles, *Language and Human Nature* (Minneapolis: University of Minnesota Press, 1985), 214.
19 See Pierre Feyereisen and Jacques-Dominique de Lannoy, *Psychologie du geste* (Bruxelles: Pierre Mardaga, 1985), 156.
20 Georg Simmel, "Aesthetic Significance of the Face," in *Essays on Sociology, Philosophy and Aesthetics,* ed. Kurt Wolff (New York: Harper Torchbooks, 1959), 278.
21 Sarles, *Language and Human Nature,* 216–17.
22 Iván Fónagy, *La vive voix: Essais de psycho-phonétique* (Paris: Payot, 1983), 116.
23 Ibid., 149. See also Iván Fónagy, "Des fonctions de l'intonation. Essai de synthèse," *Flambeau* 29 (2003): 1–20.
24 See Piaget, *Play, Dreams and Imitation,* 19–20.
25 See Iván Fónagy, "Emotions, Voice and Music," in *Research Aspects on Singing,* ed. Johan Sundberg (Stockholm: Royal Swedish Academy of Music, 1981), 74.
26 See René A. Spitz, *No and Yes: On the Genesis of Human Communication* (New York: International Universities Press, 1957), 40–43.
27 See Wolfgang Prinz, "Ideo-Motor Action," in *Perspectives on Perception and Action,* ed. Herbert Heuer and Andries F. Sanders (Hillsdale, NJ: Lawrence Erlbaum Associates, 1987), 49–53.
28 On the mimetic tendency or impulse, see Hans Prinzhorn, *Artistry of the Mentally Ill: A Contribution to the Psychology and Psychopathology of Configuration,* trans. Eric von Brockdorff (New York: Springer-Verlag, 1962), 23–26.
29 Kurt. Koffka, *The Growth of the Mind: An Introduction to Child Psychology* (New Brunswick, NJ: Transaction Books, 1980), 316–17.
30 See also Meltzoff and Moore, "Infant's Understanding of People and Things," 52–54.
31 David Abercrombie, "A Phonetician's View of Verse Structure," in *Studies in Phonetic and Linguistics* (London: Oxford University Press, 1971), 19.
32 See Buytendijk, *Le football,* 21–22.

33 Rudolf Arnheim, "Sculpture: The Nature of a Medium," in *To the Rescue of Art: Twenty-Six Essays* (Berkeley: University of California Press, 1992), 82–91.
34 See also Susan Stewart, "Prologue: From the Museum of Touch," in *Material Memories: Design and Evocation*, eds. Marius Kwint, Christopher Breward and Jeremy Aynsley (Oxford: Berg, 1999), 17–36.
35 Jacques Lecoq, with Jean-Gabriel Carasso and Jean-Claude Lallias, *The Moving Body: Teaching Creative Theatre*, trans. David Bradby (New York: Routledge, 2001), 47.
36 Rudolf Arnheim, "Art among the Objects," in *To the Rescue of Art: Twenty-Six Essays* (Berkeley: University of California Press, 1992), 7–14.
37 Piaget, *Play, Dreams and Imitation in Childhood*, 81. See also Konrad Lorenz, *Behind the Mirror: A Search for a Natural History of Human Knowledge*, trans. Ronald Taylor (New York: Harcourt Brace Jovanovich, 1977), 151–56.
38 Piaget, *Play, Dreams and Imitation in Childhood*, 70–71.
39 Merleau-Ponty, *Phenomenology of Perception*, 185. On the body's "comprehensive power," see also Buytendijk, *Traité de psychologie animale*, 325–43.
40 Ibid., 352.
41 Maurice Merleau-Ponty, *Consciousness and Language Acquisition*, trans. Hugh J. Silvermann (Evanston: Northwestern University Press, 1973), 36.
42 Helmuth Plessner, "Zur Anthropologie der Nachahmung," in *Ausdruck und menschliche Natur*, vol. 7 of *Gesammelte Schriften* (Frankfurt am Main: Suhrkamp Verlag, 1982), 391–98. Plessner makes a clear distinction between imitation and vital motor response. The former requires the gradual learning of movements and the ability to objectify the body. In the absence of such an ability, we are unable to consider our own bodily schema and the interchangeability of motor performances.
43 Janet Lynn Roseman, *Dance Masters: Interviews with Legends of Dance* (New York: Routledge, 2001), 45.
44 On the body's sympathetic awareness, see John Martin, *The Dance in Theory* (1939), e-book reprint (Highstown, NJ: Princeton Book Company, 2004), 47–55.
45 On the phenomenon of inward attitude, see Jürg Zutt, "Die innere Haltung," *Auf dem Wege zu einer anthropologischen Psychiatrie: Gesammelte Aufsätze* (Berlin: Springer-Verlag, 1963), 1–88.
46 Peter L. Berger, *Invitation to Sociology: A Humanistic Perspective* (Garden City, NY: Anchor Books, 1963), 95.
47 Zutt, "Die innere Haltung," 3–4.

48 See Lecoq, *The Moving Body*, 66–90.

49 Max Scheler, *On the Eternal in Man*, trans. Bernard Noble (New York: Harper & Brothers, 1960), 265.

50 Berger, *Invitation to Sociology*, 96.

51 Ibid., 98.

52 See Hermann H. Spitz, *Nonconscious Movements: From Mystical Messages to Facilitated Communication* (Manwah, NJ: Lawrence Erlbaum Associates, 1997). On the various ideo-motor based practises, see Ray Hyman, "The Mischief Making of Ideomotor Action," *The Scientific Review of Alternative Medicine* 3, no. 2 (1999): 34–43.

53 Alfred North Whitehead, *Modes of Thought* (New York: Free Press, 1968), 36.

54 Peter Brook, *The Empty Space* (Harmondsworth: Penguin, 1968), 122.

55 William James, "The Gospel of Relaxation," in *Writings 1878 – 1899*, ed. Gerald E. Meyers (New York: The Library of America, 1992), 835.

56 See also Luigi Bonpensiere, *New Pathways to Piano Technique: A Study of the Relations between Mind and Body with Special Reference to Piano Playing* (New York: Philosophical Library, 1967).

57 See my article, "Le jeu rituel: Pour une phénoménologie de la mémoire corporelle," *Études phénoménologiques* 36 (2002): 97–118.

58 Jürgen Habermas, "Walter Benjamin: Consciousness-Raising or Rescuing Critique (1972)," in *Philosophical-Political Profiles*, trans. Frederick G. Lawrence (Cambridge, MA: MIT Press, 1983), 147.

59 Ulrich Schwartz, "Walter Benjamin: Mimesis und Erfahrung," in *Philosophie der Gegenwart*, vol. 6 of *Grundprobleme der grossen Philosophen*, ed. Josef Speck (Göttingen: Vandenhoeck & Ruprecht, 1984), 46.

60 See Harmut Böhme and Gernot Böhme, *Das Andere der Vernunft: Zur Entwicklung von Rationalitätsstrukturen am Beispiel Kants* (Frankfurt am Main: Suhrkamp Verlag, 1983), 277–81; Christine Bernd, *Bewegung und Theater: Lernen durch Verkörpern* (Frankfurt am Main: AFRA Verlag, 1988), 76–85.

chapter 5

1 Nicolas Abraham, *Rhythms: On the Work, Translation, and Psychoanalysis*, trans. Benjamin Thigpen and Nicholas T. Rand (Stanford: Stanford University Press, 1995), 79.

2 Manfred Pohlen, "Über die Beziehung zwischen rhythmischer Einstimmung und frühzeitiger Differenzierung des Gehörsinns bei der Entstehung des Ich und der Sprache," *Jahrbuch für Psychologie, Psychotherapie und medizinische Anthropologie* 17 (1969): 288.

3 Edward T. Hall, "Rhythm and Body Movement," in *Beyond Culture* (New York: Anchor Press, 1976), 75.

4 See Albert E. Scheflen, "Comments on the Significance of Interaction Rhythms," in *Interaction Rhythms: Periodicity in Communicative Behavior*, ed. Martha Davis (New York: Human Sciences Press, 1982), 17.
5 See Campbell, "The Conversational Waltz," 229–46.
6 Schutz, "Making Music Together," 176.
7 See F.J.J. Buytendijk, *Phénoménologie de la rencontre*, trans. Jean Knapp (Paris: Desclée de Brouwer, 1952).
8 See Carl E. Seashore, *Psychology of Music* (New York: Dover, 1967), 138–48.
9 Merleau-Ponty, *Phenomenology of Perception*, 146.
10 Kurt Goldstein, *The Organism: A Holistic Approach to Biology Derived from Pathological Data in Man* (New York: Zone Books, 2000), 283. Seashore also insists that "all rhythm is primarily a projection of personality. The rhythm is what I am." *Psychology of Music*, 139.
11 See Leonard C. Feldstein, "The Human Body as Rhythm and Symbol: A Study in Practical Hermeneutics," *The Journal of Medicine and Philosophy* 1 (1976): 141–43.
12 Oliver Sacks, *A Leg to Stand On*, 148.
13 Ibid., 149.
14 For an excellent overview, see Peter Röthig, "Betrachtungen zur Körper- und Bewegungsästhetik," 88–95.
15 See, for instance, Buytendijk, *Allgemeine Theorie*, 357–64. See also W.J.M. Dekkers, "The Lived Body as Aesthetic Object in Anthropological Medicine," *Medicine, Health Care and Philosophy* 2 (1999): 122.
16 See Raimund Sobotka, *Formgesetze der Bewegungen im Sport* (Schorndorf: Verlag Karl Hofmann, 1974), 109–22.
17 See Peter Röthig, "Bewegungsgestaltung and ästhetische Erziehung im Sport," in *Facetten der Sportpädagogik*, ed. Robert Prohl (Schorndorf: Verlag Karl Hofmann, 1993), 13–22.
18 I borrow this expression from Susanne K. Langer. See her *Philosophy in a New Key: A Study in the Symbolism of Reason, Rite, and Art*, 3rd ed. (Cambridge, MA: Harvard University Press, 1957), 293.
19 See Paul Guillaume, *La formation des habitudes*, new ed. (Paris: PUF, 1968), 104–9.
20 Ibid., 105.
21 See Buytendijk, *Allgemeine Theorie*, 280–86.
22 See Peter Röthig, *Rhythmus und Bewegung: Eine Analyse aus der Sicht der Leibeserziehung*, 2nd ed. (Schorndorf: Verlag Karl Hofmann, 1984); Peter Röthig, "Zur Theorie des Rhythmus," in *Grundlagen und Perspektiven ästhetischer und rhythmischer Bewegungserziehung*, eds. Eva Bannmüller and Peter Röthig (Stuttgart: Ernst Klett Verlag, 1990), 51–71.

23 Paul Souriau, *Aesthetics of Movement*, trans. and ed. Manon Souriau (Amherst: University of Massachusetts Press, 1983), 25.

24 Röthig, *Rhythmus und Bewegung*, 93. See also Inge Heuser, "Rhythmus als Ausdruck des Lebendigen," in *Beiträge zur Theorie und Lehre vom Rhybmus*, ed. Peter Röthig (Schorndorf: Verlag Karl Hofmann, 1966) 122–36.

25 Lewis Mumford, *Art and Technics* (New York: Columbia University Press, 1960), 62.

26 The expression is taken from the beautiful essay of Henri Focillon, "In Praise of Hands," in *The Life of Forms in Art*, trans. Charles Beecher Hogan and George Kubler (New York: Zone Books, 1992), 170.

27 Straus, "The Forms of Spatiality," 29.

28 Ibid., 34–35.

29 Ibid., 26. The dancer Eric Hawkins declared: "Isadora Duncan was the first dancer in the West to intuit a kinesiological truth: that human movement starts in the spine and pelvis, not in the extremities – the legs and arms. That is: human movement, when it obeys the nature of its functioning, when it is not distorted by erroneous concepts of the mind, starts in the body's center of gravity and then – in correct sequence – flows into the extremities." "Pure Poetry," in *The Modern Dance: Seven Statements of Belief*, ed. Selma Jeanne Cohen (Middletown, CT: Wesleyan University Press, 1969), 41.

30 Rudolf Arnheim, "Concerning Dance," in *Toward a Psychology of Art: Collected Essays* (Berkeley: University of California Press, 1966), 264.

31 Judith Lynne Hanna, *To Dance is Human: A Theory of Nonverbal Communication* (Chicago: University of Chicago Press, 1987), 34.

32 Gehlen, *Der Mensch*, 190–92.

33 Ursula Fritsch, "Tanz 'stellt nicht dar, sondern macht wirklich': Ästhetischer Erziehung als Ausbildung tänzerischer Sprachfähigkeit," in *Grundlagen und Perspektiven ästhetischer und rhythmischer Bewegungserziehung*, eds. Eva Bannmüller and Peter Röthig (Stuttgart: Ernst Klett Verlag, 1990), 110.

34 Maxine Sheets-Johnson, *The Phenomenology of Dance* (New York: Books for Libraries, 1980), 102.

35 Jacqueline Lessschaeve, *The Dancer and the Dance: Conversation with Merce Cunningham* (New York: Marion Boyars, 1999), 68. Paul Taylor reinforces this view by stating that dancers, while executing some movement sequences, feel a "peculiar kind of muscle logic." Their movement is "organic" when the new elements naturally weave into the "physical logic of the phrase." See his "Down with Choreography," in *The Modern Dance: Seven Statements of Belief*, ed. Selma Jeanne Cohen (Middletown, CT: Wesleyan University Press, 1969), 95.

36 Gehlen, *Der Mensch*, 222–27.
37 See Roderyk Lange, *The Nature of Dance: An Anthropological Perspective* (London: Macdonald & Evans, 1975), 36.
38 Paul Valéry, "Philosophy of the Dance," in *Aesthetics*, trans. Ralph Manheim (New York: Pantheon Books, 1964), 208.
39 Merce Cunningham, "The Impermanent Art (1952)," in *Merce Cunningham: Fifty Years*, chronicle and commentary by David Vaughan and edited by Melissa Harris (New York: Aperture, 1997), 86.
40 Valéry, "Philosophy of Dance," 204.
41 Paul Valéry, "Degas, Dance, Drawing," in *Degas, Manet, Morisot*, trans. David Paul (New York: Pantheon Books, 1960),15.
42 See also Ursula Fritsch, *Tanz, Bewegung, Gesellschaft: Verluste und Chancen symbolisch-expressiven Bewegens* (Frankfurt am Main: AFRA Verlag, 1988), 41–48.
43 Seashore, *Psychology of Music*, 142.
44 Valéry, *Cahiers*, 1279, 1283. See also Bernhard Waldenfels, "Vom Rhythmus der Sinnen," in *Sinnesschwellen: Studien zur Phänomenologie des Fremden* (Frankfurt am Main: Suhrkamp Verlag, 1999), 79–83.
45 Peter Röthig, "Bewegung – Rhythmus – Gestaltung: Zu Problemen gymnastischer Kategorien," in *Gymnastik: Ein Beitrag zur Bewegungskultur unserer Gesellschaft*, eds. Klaus-Jürgen Gutsche and Hans Jochen Medau (Schorndorf: Verlag Karl Hofmann, 1989), 42–45.
46 See Lange, *The Nature of Dance*, 33.
47 Souriau, *Aesthetics of Movement*, 23.
48 See Hermann Schmitz, *Leib and Gefühl: Materialen zu einer philosophischen Therapeutik*, eds. Hermann Gausebeck and Gerhard Risch (Padeborn: Junfermann-Verlag, 1989), 121, 141–42; Buytendijk, *Wesen und Sinn des Spiels*, 73–75.
49 See Gehlen, *Der Mensch*, 144.
50 Buytendijk, *Wesen und Sinn des Spiels*, 79.

chapter 6

1 Merleau-Ponty, *Phenomenology of Perception*, 181.
2 Gabriel Marcel, "Leibliche Begegnung: Notizen aus einem gemeinsamen Gedankengang," in *Leiblichkeit: Philosophische, gesellschaftliche und therapeutische Perspektiven*, ed. Hilarion Petzold (Padeborn: Junfermann-Verlag, 1986), 34–39.
3 Merleau-Ponty, *Phenomenology of Perception*, 198.
4 See Haeffner, *Philosophische Anthropologie*, 110–11.
5 Ricoeur, *Freedom and Nature*, 280.

6 Buytendijk, *Prolegomena*, 62.
7 Arnold Gehlen, "Vom Wesen der Erfahrung," in *Anthropologische und sozialpsychologische Untersuchungen* (Reinbek: Rowohlt Verlag, 1993), 29–31.
8 Jerome S. Bruner, "Modalities of Memory," in *The Pathology of Memory*, eds. George A. Talland and Nancy C. Waugh (New York: Academic Press, 1969), 254. See also Thomas Fuchs, "Das Gedächtnis des Leibes," *Phänomenologische Forschungen* 5 (2000): 71–76; Paul Brockelmann, "Of Memory and Things Past," *International Philosophical Quarterly* 15 (1975): 314–16.
9 Merleau-Ponty, *Phenomenology of Perception*, 142–47; Buytendijk, *Allgemeine Theorie*, 266–68 and *Prolegomena*, 34–39.
10 Straus, *The Primary World of Senses*, 256–59.
11 Merleau-Ponty, *Phenomenology of Perception*, 143.
12 Bruner, "Modalities of Memory," 254.
13 Buytendijk, *Allgemeine Theorie*, 272.
14 Ricoeur, *Freedom and Nature*, 283.
15 See Herbert Plügge, *Vom Spielraum des Leibes: Klinisch-phänomenologische Erwägungen über "Körperschema" und "Phantomglied"* (Salzburg: Otto Müller Verlag, 1970), 70.
16 Merleau-Ponty, *Phenomenology of Perception*, 130. See also Antonio Mazzù, "Syntaxe motrice et stylistique corporelle: Réflexions à propos du schématisme corporel chez le premier Merleau-Ponty," *Revue philosophique de Louvain* 99 (2001): 46–72.
17 See Bruner, "Modalities of Memory," 257–58.
18 Samuel Butler, *Life and Habit* (London: Jonathan Cape, 1921), 1–77.
19 See also Samuel Butler, *The Note-Books of Samuel Butler*, ed. Henry Festing Jones (London: A. C. Fifield, 1913), 47–55; Michael Polanyi, *Personal Knowledge: Towards a Post-Critical Philosophy* (Chicago: University of Chicago Press, 1962), 55–57.
20 See György Sándor, *On Piano Playing: Motion, Sound and Expression* (New York: Schirmer Books, 1995), 188.
21 Merleau-Ponty, *Phenomenology of Perception*, 145.
22 Ibid., 143.
23 Maurice Merleau-Ponty, *Phénoménologie de la perception* (Paris; Gallimard, 1945), 169. We refer here to the original text, because writing "a certain type" in the translation is incorrect.
24 Ricoeur, *Freedom and Nature*, 287.
25 Merleau-Ponty, *Phenomenology of Perception*, 153. See also Linda Singer, "Merleau-Ponty on the Concept of Style," *Man and World* 14 (1981): 153–63.
26 Merleau-Ponty, *Phenomenology of Perception*, 151.

27 See van den Berg, *A Different Existence*, 62.
28 Buytendijk, *Mensch und Tier*, 46.
29 Merleau-Ponty, *Phenomenology of Perception*, 238.
30 See D. J. van Lennep, "The Psychology of Driving a Car," in *Phenomenological Psychology: The Dutch School*, ed. Joseph J. Kockelmans (Dordrecht: Martinus Nijhoff, 1987), 217–27.
31 Maurice Merleau-Ponty, "Indirect Language and the Voices of Silence," in *Signs*, trans. Richard C. McClearly (Evanston: Northwestern University Press, 1964), 65.
32 Ricoeur, *Freedom and Nature*, 287.
33 Merleau-Ponty, "Man and Adversity," 235.
34 Ricoeur, *Freedom and Nature*, 289.
35 Erik H. Erikson, "Ontogeny of Ritualization in Man," *Philosophical Transactions of the Royal Society of London: Series B. Biological Sciences* 251 (1966): 337–49.
36 John Dewey, *Human Nature and Conduct: An Introduction to Social Psychology* (New York: Modern Library, 1957), 128.
37 Ricoeur, *Freedom and Nature*, 305.
38 See van Lennep, "The Psychology of Driving a Car," 222.
39 See Wilfried Ennenbach, *Bild und Mitbewegung* (Köln: bps-Verlag, 1991), 88–94.
40 See Goldstein, *The Organism*, 34.
41 Hubertus Tellenbach, "The Education of Medical Student," in *The Moral Sense in the Communal Significance of Life*, vol. 20 of *Analecta Husserliana*, ed. Anna-Teresa Tymieniecka (Dordrecht: D. Riedel, 1986), 181.

chapter 7

1 Melchior Palágyi, *Wahrnehmungslehre*, vol. 2 of *Ausgewählte Werke* (Leipzig: Johann Ambrosius Barth, 1925), 69–105; *Naturphilosophische Vorlesungen: Über die Grundprobleme des Bewusstseins und des Lebens*, vol. 1 of *Ausgewählte Werke* (Leipzig: Johann Ambrosius Barth, 1924). For an overview of Palágyi's ideas in English, see W. R. Boyce Gibson, "The Philosophy of Melchior Palágyi," *Journal of Philosophical Studies* 3 (1928): 15–28, 157–72.
2 Palágyi, *Naturphilosophische Vorlesungen*, 163.
3 Palágyi, *Wahrnehmungslehre*, 79.
4 The Japanese philosopher Ichikawa Hiroshi voiced a similar view about tactile perception: "When we grasp a stone, we sketch its possible shapes and respond to both to its actual and possible shapes.... To explain it in reverse, we elicit the stone's response by grasping and posing questions to it." The quote is taken from Shigenori Nagatomo's article,

"Two Contemporary Japanese Views of the Body: Ichikawa Hiroshi and Yuasa Yasuo," in *Self as Body in Asian Theory and Practice*, ed. Thomas P. Kasulis (Albany: State University of New York Press, 1993), 325.

5 See Palágyi, *Wahrnehmungslehre*, 97.

6 See Palágyi, *Naturphilosophische Vorlesungen*, 165.

7 Ibid., 160–63.

8 Ibid., 169–70.

9 Ibid., 226.

10 Jean Cocteau, *Journal d'un inconnu* (Paris: Bernard Grasset, 1953), 165. This passage has been analyzed in depth by Jan Hendrick van den Berg, *The Changing Nature of Man: Introduction to a Historical Psychology (Metabletica)*, trans. H. F. Croes (New York: W. W. Norton, 1983), 211–17.

11 Buytendijk, "Some Aspects of Touch," 114.

12 Merleau-Ponty, *Phenomenology of Perception*, 145–46. The tactile and vibratory aspects of music-making are central for the African artistic consciousness. See Robert Kauffman, "Tactility as an Aesthetic Consideration in African Music," in *The Performing Arts: Music and Dance*, eds. John Blacking and Joann W. Kealiinohomoku (The Hague: Mouton, 1979), 251–53.

13 Palágyi, *Wahrnehmungslehre*, 94.

14 Gehlen, *Der Mensch*, 185–86.

15 Focillon, "In Praise of Hands," 162–63.

16 Green (with Gallwey), *The Inner Game of Music*, 99–100.

17 See Anton Ehrenzweig, *The Hidden Order of Art: A Study in the Psychology of Artistic Imagination* (Berkeley: University of California Press, 1995), 43–44.

18 Palágyi, *Naturphilosophische Vorlesungen*, 130.

19 Menyhért Palágyi, *Székely Bertalan és a festészet aesthetikája* (Bertalan Székely and the Aesthetics of Painting) (Budapest, Hoffmann & Vastagh, 1910), 28.

20 I recently delighted in the tactile spontaneity and expression of the Khmer sculptors who carved the astounding bas-reliefs of the Angkorian temples.

21 Focillon, "In Praise of Hands," 180.

22 Rudolf Arnheim, "Art for the Blind," in *To the Rescue of Art: Twenty-Six Essays* (Berkeley: University of California Press, 1992), 139–40.

23 Gert Selle, *Gebrauch der Sinne: Eine kunstpädagogische Praxis* (Reinbek: Rowohlt Verlag, 1988), 225–34.

24 Ibid., 230.

25 Ibid., 234.

26 Focillon, "In Praise of Hands," 180.

27 Géza Révész, "La fonction sociologique de la main humaine et de la main animale," *Journal de psychologie normale et pathologique* 35 (1938): 45–46.

28 See also Alexander Gosztonyi, *Der Mensch in der modernen Malerei: Versuche zur Philosophie des Schöpferischen* (Munich: Verlag C. H. Beck, 1970), 169–75; Hermann Schmitz, *Der Leib im Spiegel der Kunst*, vol. II, 2 of *System der Philosophie* (Bonn: Bouvier Verlag, 1966), 69–81.

29 Alfred North Whitehead, *The Aims of Education and Other Essays* (New York: Free Press, 1968), 50–51.

conclusion

1 Merleau-Ponty, "Indirect Language and the Voices of Silence," 83.

2 See, for instance, Mihaly Csikszentmihalyi, *Beyond Boredom and Anxiety: The Experience of Play in Work and Games*, 2nd ed. (San Francisco: Josey Bass, 2000).

3 G. Böhme, *Anthropologie in pragmatischer Hinsicht*, 131–37.

4 See Jan Hendrik van den Berg, *Der Kranke: Ein Kapitel medizinischer Psychologie für jedermann* (Göttingen: Vandenhoeck & Ruprecht, 1961), 30–34.

5 See Karlfried Graf Dürckheim, *Hara: The Vital Centre of Man*, trans. Sylvia-Monica von Kospoth (London: Mandela Book, 1988), 152–64.

6 Eugen Herrigel, *Zen in the Art of Archery*, trans. R.F.C. Hull (New York: Vintage Books, 1989), 40.

7 Much can be learned from the oriental understanding of the body and bodily practices. See Ichiro Yamaguchi, *Ki als leibhaftige Vernunft: Beitrag zur interkulturellen Phänomenologie der Leiblichkeit* (Munich: Wilhelm Fink Verlag, 1997).

8 G. Böhme, *Anthropologie in pragmatischer Hinsicht*, 137. See also the essay of Jan Linschoten, "Aspects of the Sexual Incarnation. An Inquiry Concerning the Meaning of the Body in Sexual Encounter," in *Phenomenological Psychology: The Dutch School*, ed. Joseph J. Kockelmans (Dordrecht: Martinus Nijhoff, 1987), 149–94.

9 Jan Kott, "The Memory of the Body," in *The Memory of the Body: Essays on Theater and Death* (Evanston: Northwestern University Press, 1992), 115. Horkheimer and Adorno also speak of "original unity" and argue that "sex represents the body in its purest state." *Dialectic of Enlightenment*, 235.

10 Helmuth Plessner, *Laughing and Crying: A Study of the Limits of Human Behavior*, trans. James S. Churchill and Marjorie Grene (Evanston: Northwestern University Press, 1970).

11 Ibid., 148.

12 Straus, "The Forms of Spatiality," 30–37.

13 Heinrich von Kleist, "Reflection: A Paradox," in *Selected Writings*, ed. and trans. David Constantine (London: J. M. Dent, 1997), 410.
14 Walter Benjamin, "Die glückliche Hand: Eine Unterhaltung über das Spiel," in *Gesammelte Schriften*, vol. IV, 2, ed. Tillman Rexroth (Frankfurt am Main: Suhrkamp Verlag, 1972), 776.
15 James, "The Gospel of Relaxation," 836–37.
16 Aldous Huxley, "Knowledge and Understanding," in *Adonis and the Alphabet and Other Essays* (London: Chatto & Windus, 1956), 64–65.
17 See Hans-Eduard Hengstenberg, *Philosophische Anthropologie*, 2nd ed. (Stuttgart: Verlag W. Kohlhammer, 1957), 263–66; "Phenomenology and Metaphysics of the Human Body," *International Philosophical Quarterly* 3 (1963): 165–200.
18 Hengstenberg, *Philosophische Anthropologie*, 264.
19 See Jürg Zutt, "Der Leib der Tiere," in *Auf dem Wege zu einer anthropologischen Psychiatrie. Gesammelte Aufsätze* (Berlin: Springer-Verlag, 1963), 460–61.

Bibliography

Abercrombie, David. "Conversation and Spoken Prose." In *Studies in Phonetic and Linguistics*, 1–9. London: Oxford University Press, 1971.

———. "A Phonetician's View of Verse Structure." In *Studies in Phonetic and Linguistics*, 16–25. London: Oxford University Press, 1971.

Arnheim, Rudolf. "Concerning Dance." In *Toward a Psychology of Art: Collected Essays*, 261–65. Berkeley: University of California Press, 1966.

———. *Visual Thinking*. Berkeley: University of California Press, 1969.

———. "Art Among the Objects." In *To the Rescue of Art: Twenty-Six Essays*, 7–14. Berkeley: University of California Press, 1992.

———. "Perceptual Aspects of Art for the Blind." In *To the Rescue of Art: Twenty-Six Essays*, 133–43. Berkeley: University of California Press, 1992.

———. "Sculpture: *The Nature of a Medium*." In *To the Rescue of Art: Twenty-Six Essays*, 82–91. Berkeley: University of California Press, 1992.

Abraham, Nicolas. *Rhythms: On the Work, Translation, and Psychoanalysis.* Translated by Benjamin Thigpen and Nicholas T. Rand. Stanford: Stanford University Press, 1995.

Bachmann, Klaus. "Körper-Intelligenz: Das motorische Wunder." *Magazin GEO* 8/August 1999, 14–34.

Barba, Eugenio. *The Paper Canoe: A Guide to Theatre Anthropology.* Translated by Richard Fowler. London: Routledge, 1995.

———. "The Dilated Body." In *A Dictionary of Theatre Anthropology: The Secret Art of the Performer*, edited by Eugenio Barba and Nicola Savarese, 54–63. London: Routledge, 1995.

Barbaras, Renaud. "Affectivity and Movement: The Sense of Sensing in Erwin Straus." *Phenomenology and the Cognitive Sciences* 3 (2004): 215–28.

Bateson, Gregory. "Conscious Purpose versus Nature." In *Steps to an Ecology of Mind: Collected Essays in Anthropology, Psychiatry, Evolution, and Epistemology*, 432–45. Northvale, NJ: Jason Aronson, 1987.

Benjamin, Walter. "Die glückliche Hand: Eine Unterhaltung über das Spiel." In *Gesammelte Schriften*. Vol. IV, 2. Edited by Tillman Rexroth, 771–77. Frankfurt am Main: Suhrkamp Verlag, 1972.

———. "On the Mimetic Faculty." In *Reflections*, edited by Peter Demetz, translated by Edmund Jephcott, 333–36. New York: Schocken Books, 1986,

Berg, Jan Hendrik van den. *The Changing Nature of Man: Introduction to a Historical Psychology (Metabletica).* Translated by H. F. Croes. New York: W. W. Norton, 1983.

———. *Der Kranke: Ein Kapitel medizinischer Psychologie für jedermann.* Göttingen: Vandenhoeck & Ruprecht, 1961.

———. *A Different Existence: Principles of Phenomenological Psychopathology.* Pittsburgh: Duquesne University Press, 1995.

Berger, Peter L. *Invitation to Sociology: A Humanistic Perspective.* Garden City, NY: Anchor Books, 1963.

Bernd, Christine. *Bewegung und Theater: Lernen durch Verkörpern.* Frankfurt am Main: AFRA Verlag, 1988.

Blacking, John. "Towards an Anthropology of the Body." In *The Anthropology of the Body*, edited by John Blacking, 1–28. London: Academic Press, 1977.

———. "The Biology of Music-Making." In *Ethnomusicology*. Vol. 1 of *The New Grove Dictionary of Music and Musicians*, edited by Helen Myers, 301–14. London: Macmillan Press, 1992.

Bloch, Ernst. *The Principle of Hope.* Translated by Neville Place, Stephen Place and Paul Knight. Cambridge, MA: MIT Press, 1986.

Bloomer, Kent C., and Charles W. Moore. *Body, Memory, and Architecture.* New Haven: Yale University Press, 1977.

Böhme, Gernot. "Atmosphäre als Grundbegriff einer neuen Ästhetik." In *Atmosphäre: Essays zur neuen Ästhetik*, 21–48. Frankfurt am Main: Suhrkamp Verlag, 1995.

———. *Anthropologie in pragmatischer Hinsicht: Darmstädter Vorlesungen*. Frankfurt am Main: Suhrkamp Verlag, 1985.

Böhme, Harmut, and Gernot Böhme. *Das Andere der Vernunft: Zur Entwicklung von Rationalitätsstrukturen am Beispiel Kants*. Frankfurt am Main: Suhrkamp Verlag, 1983.

Bollnow, Otto Friedrich. *Die pädagogische Atmosphäre: Untersuchungen über die gefühlsmäßigen zwischenmenschlichen Voraussetzungen der Erziehung*. Essen: Verlag Die Blaue Eule, 2001.

Bonpensiere, Luigi. *New Pathways to Piano Technique: A Study of the Relations between Mind and Body with Special Reference to Piano Playing*. New York: Philosophical Library, 1967.

Boorstin, Daniel J. *The Image or What Happened to the American Dream*. Harmondsworth: Penguin Books, 1961.

Borgmann, Albert. *Technology and the Character of Contemporary Life: A Philosophical Inquiry*. Chicago: University of Chicago Press, 1984.

———. *Crossing the Postmodern Divide*. Chicago: University of Chicago Press, 1992.

Boss, Medard. *Existential Foundations of Medicine and Psychology*. Translated by Stephen Conway and Anne Cleaves. New York: Jason Aronson, 1984.

Boyce Gibson, W. R. "The Philosophy of Melchior Palágyi." *Journal of Philosophical Studies* 3 (1928): 15–28, 157–72.

Brockelmann, Paul. "Of Memory and Things Past." *International Philosophical Quarterly* 15 (1975): 309–25.

Brook, Peter. *The Empty Space*. Harmondsworth: Penguin, 1968.

Bruaire, Claude. *Philosophie du corps*. Paris: Seuil, 1968.

Bruner, Jerome S. "Modalities of Memory." In *The Pathology of Memory*, edited by George A. Talland and Nancy C. Waugh, 253–59. New York: Academic Press, 1969.

Buckland, Stephen. "Ritual, Bodies and 'Cultural Memory'." In *Liturgy and the Body*, edited by Louis-Marie Chauvet and François Kabasele Lumbala, 49–56. London: SCM Press, 1995.

Butler, Samuel. *The Note-Books of Samuel Butler*. Edited by Henry Festing Jones. London: A. C. Fifield, 1913.

———. *Life and Habit*. London: Jonathan Cape, 1921.

Buytendijk, F.J.J. *Le football: Une étude psychologique*. Paris: Desclée de Brouwer, 1952.

———. *Phénoménologie de la rencontre*. Translated by Jean Knapp. Paris: Desclée de Brouwer, 1952.

———. *Traité de psychologie animale*. Translated by A. Frank-Duquesne. Paris: PUF, 1952.
———. *Allgemeine Theorie der menschlichen Haltung und Bewegung*. Berlin: Springer-Verlag, 1956.
———. "Das Menschliche der menschlichen Bewegung." In *Das Menschliche: Wege zu seinem Verständnis*, 170–88. Stuttgart: Koehler Verlag, 1958.
———. *Mensch und Tier: Ein Beitrag zur vergleichenden Psychologie*. Reinbek: Rowohlt Verlag, 1958.
———. *Prolegomena einer anthropologischen Physiologie*. Salzburg: Otto Müller Verlag, 1967.
———. "L'objectivité des choses et l'expressivité des formes." *Psychiatria, Neurologia, Neurochirurgica* 73 (1970): 427–31.
———. "Some Aspects of Touch." *Journal of Phenomenological Psychology* 1 (1970): 99–124.
———. *Wesen und Sinn des Spiels: Das Spielen des Menschen und der Tiere als Erscheinungsform der Lebenstriebe* (1933). Reprint. New York: Arno Press, 1976.
Campbell, Jeremy. "The Conversational Waltz." In *Winston Churchill's Afternoon Nap*, 229–46. New York: Simon and Schuster, 1986.
Cannon, Walter B. *The Wisdom of the Body*. 2nd ed. Magnolia, MA: Peter Smith, 1978.
Cassirer, Ernst. *The Philosophy of Symbolic Forms*. Vol. 1. *Language*. Translated by Ralph Manheim. New Haven: Yale University Press, 1955.
Chekhov, Michael. *To the Actor*. London: Routledge, 2002.
Clarke, Eric F. "Generative Principles in Music Performance." In *Generative Processes in Music: The Psychology of Performance, Improvisation, and Composition*, edited by John A. Sloboda, 1–26. Oxford: Clarendon Press, 1988.
Cocteau, Jean. *Journal d'un inconnu*. Paris: Bernard Grasset, 1953.
Christian, Paul. "Möglichkeiten und Grenzen einer naturwissenschaftlichen Betrachtung der menschlichen Bewegung," *Jahrbuch für Psychologie und Psychopathologie* 4 (1956): 346–56.
———. "Vom Wertbewusstsein im Tun. Ein Beitrag zur Psychophysik der Willkürbewegung." In Frederik J. J. Buytendijk, Paul Christian and Herbert Plügge, *Über die menschliche Bwegung als Einheit von Natur und* Geist, 19–44. Schorndorf: Verlag Karl Hofmann, 1963.
Csepregi, Gabor. "Du tact." In *Science et Esprit* 47 (1995): 329–33.
———. "Le jeu rituel: Pour une phénoménologie de la mémoire corporelle." *Études phénoménologiques* 36 (2002): 97–118.
———. ed. *Sagesse du corps*. Alymer: Éditions du Scribe, 2001.
Csikszentmihalyi, Mihaly. *Beyond Boredom and Anxiety: The Experience of Play in Work and Games*. 2nd ed. San Francisco: Josey Bass, 2000.

Cunningham, Merce. "The Impermanent Art (1952)." In *Merce Cunningham: Fifty Years*. Chronicle and commentary by David Vaughan and edited by Melissa Harris, 86–87. New York: Aperture, 1997.
Dekkers, W.J.M. "The Lived Body as Aesthetic Object in Anthropological Medicine." *Medicine, Health Care and Philosophy* 2 (1999): 117–28.
De Koninck, Thomas. *Philosophie de l'éducation: Essai sur le devenir humain*. Paris: PUF, 2004.
Deutsch, Eliot. *Personhood, Creativity and Freedom*. Honolulu: University of Hawaii Press, 1982.
Dewey, John. *Human Nature and Conduct: An Introduction to Social Psychology*. New York: Modern Library, 1957.
Dürckheim, Karlfried Graf. *Hara: The Vital Centre of Man*. Translated by Sylvia-Monica von Kospoth. London: Mandela Book, 1988.
Ehrenzweig, Anton. *The Hidden Order of Art: A Study in the Psychology of Artistic Imagination*. Berkeley: University of California Press, 1995.
Ennenbach, Wilfried. *Bild und Mitbewegung*. Köln: bps-Verlag, 1991.
Erikson, Erik H. "Ontogeny of Ritualization in Man." *Philosophical Transactions of the Royal Society of London: Series B. Biological Sciences* 251 (1966): 337–49.
Erikson, Joan M. *Wisdom and the Senses: The Way to Creativity*. New York: W. W. Norton, 1988.
Feldstein, Leonard C. "The Human Body as Rhythm and Symbol: A Study in Practical Hermeneutics," *The Journal of Medicine and Philosophy* 1 (1976): 136–61.
Feyereisen, Pierre, and Jacques-Dominique de Lannoy. *Psychologie du geste*. Bruxelles: Pierre Mardaga, 1985.
Focillon, Henri. "In Praise of Hands." In *The Life of Forms in Art*. Translated by Charles Beecher Hogan and George Kubler, 157–86. New York: Zone Books, 1992.
Fónagy, Iván ."Emotions, Voice and Music." In *Research Aspects on Singing*, edited by Johan Sundberg, 51–79. Stockholm: Royal Swedish Academy of Music, 1981.
———. *La vive voix. Essais de psycho-phonétique*. Paris: Payot, 1983.
———. "Des fonctions de l'intonation. Essai de synthèse." *Flambeau* 29 (2003): 1–20.
Fritsch, Ursula. *Tanz, Bewegung, Gesellschaft: Verluste und Chancen symbolisch-expressiven Bewegens*. Frankfurt am Main: AFRA Verlag, 1988.
———. "Tanz 'stellt nicht dar, sondern macht wirklich': Ästhetischer Erziehung als Ausbildung tänzerischer Sprachfähigkeit." In *Grundlagen und Perspektiven ästhetischer und rhythmischer Bewegungserziehung*, edited by Eva Bannmüller and Peter Röthig, 99–117. Stuttgart: Ernst Klett Verlag, 1990.

Fuchs, Thomas. "Das Gedächtnis des Leibes." *Phänomenologische Forschungen* 5 (2000): 71–89.

Gadamer, Hans-Georg. "The Nature of Things and the Language of Things." In *Philosophical Hermeneutics*. Translated and edited by David E. Linge, 69–81. Berkeley: University of California Press, 1977.

Gardner, Howard. *Frames of Mind: The Theory of Multiple Intelligences*. New York: Basic Books, 1993.

Gebsattel, Viktor Emil Freiherr von. "Süchtiges Verhalten im Gebiet sexueller Verirrungen." In *Prolegomena einer medizinischen Anthropologie: Ausgewählte Aufsätze*, 161–212. Berlin: Springer-Verlag, 1954.

Gehlen, Arnold. *Der Mensch: Seine Natur und seine Stellung in der Welt*. 12th ed. Wiesbaden: Akademische Verlagsgesellschaft Athenaion, 1978.

———. *Anthropologische und sozialpsychologische Untersuchungen*. Reinbek: Rowohlt Verlag, 1993.

Gilson, Etienne. *Painting and Reality*. New York: Pantheon Books, 1957.

Goldstein, Kurt. *The Organism: A Holistic Approach to Biology Derived from Pathological Data in Man*. New York: Zone Books, 2000.

Gosztonyi, Alexander. *Der Mensch in der modernen Malerei: Versuche zur Philosophie des Schöpferischen*. Munich: Verlag C. H. Beck, 1970.

Green, Barry, with W. Timothy Gallwey. *The Inner Game of Music*. New York: Doubleday, 1986.

Guillaume, Paul. *La formation des habitudes*. New ed. Paris: PUF, 1968.

Habermas, Jürgen. "Walter Benjamin: Consciousness-Raising or Rescuing Critique (1972)." In *Philosophical-Political Profiles*. Translated by Frederick G. Lawrence, 129–63. Cambridge, MA: MIT Press, 1983.

Haeffner, Gerd. *Philosophische Anthropologie*. 3rd rev. ed. Vol. 1 of *Grundkurs Philosophie*. Stuttgart: Verlag W. Kohlhammer, 2000.

Hall, Edward T. "Rhythm and Body Movement." In *Beyond Culture*, 71–84. New York: Anchor Press, 1976.

Hammer, Felix. *Leib und Geschlecht: Philosophische Perspektiven von Nietzsche bis Merleau-Ponty und phänomenologisch-systematischer Aufriss*. Bonn: Bouvier Verlag, 1974.

Hanna, Judith Lynne. *To Dance is Human: A Theory of Nonverbal Communication*. Chicago: University of Chicago Press, 1987.

Hanna, Thomas. *The Body of Life*. New York: Alfred A. Knopf, 1980.

Hartmann, Nicolaï. *Ästhetik*. 2nd ed. Berlin: Walter de Gruyter, 1966.

Hauskeller, Michael. *Atmosphären erleben: Philosophische Untersuchungen zur Sinneswahrnehmung*. Berin: Akademie Verlag 1995.

Hawkins, Eric. "Pure Poetry." In *The Modern Dance: Seven Statements of Belief*, edited by Selma Jeanne Cohen, 39–51. Middletown, CT: Wesleyan University Press, 1969.

Heckhausen, Heinz. "Entwurf einer Psychologie des Spielens." In *Das Kinderspiel*, edited by Andreas Flitner, 138–55. Munich: Piper, 1978.
Hengstenberg, Hans-Eduard. "Phenomenology and Metaphysics of the Human Body." *International Philosophical Quarterly* 3 (1963): 165–200.
———. *Philosophische Anthropologie*. 2nd ed. Stuttgart: Verlag W. Kohlhammer, 1957.
Henry, Michel. *The Essence of Manifestation*. Translated by Girard Etzkorn. The Hague: Martinus Nijhoff, 1973.
Herrigel, Eugen. *Zen in the Art of Archery*. Translated by R.F.C. Hull. New York: Vintage Books, 1989.
Heuser, Inge. "Rhythmus als Ausdruck des Lebendigen." In *Beiträge zur Theorie und Lehre vom Rhyhmus*, edited by Peter Röthig, 122–36. Schorndorf: Verlag Karl Hofmann, 1966.
Heyd, David. "Tact: Sense, Sensitivity, and Virtue." In *Inquiry* 38 (1995): 217–31.
Hiss, Tony. *The Experience of Place*. New York: Vintage Books, 1991.
Horkheimer, Max. *Eclipse of Reason*. New York: Oxford University Press, 1947.
Horkheimer, Max, and Theodor W Adorno. *Dialectic of Enlightenment: Philosophical Fragments*. Translated by John Cumming. New York: Continuum, 1989.
Hughes, Susan M., Franco Dispenza, and Gordon G. Gallup Jr. "Ratings of Voice Attractiveness Predict Sexual Behavior and Body Configuration." *Evolution and Human Behavior* 25 (2004): 295–304.
Huxley, Aldous. "The Education of an Amphibian." In *Adonis and the Alphabet and Other Essays*, 9–38. London: Chatto & Windus, 1956.
———. "Education on the Nonverbal Level," *Deadalus* 91 (1962): 279–93.
———. Foreword to *New Pathways to Piano Technique: A Study of the Relations between Mind and Body with Special Reference to Piano Playing* by Luigi Bonpensiere, v–xiii. New York: Philosophical Library, 1967.
———. "The Ego." In *The Human Situation: Lectures at Santa Barbara, 1959*. Edited by Piero Ferrucci, 129–42. London: Flamingo Modern Classic, 1994.
Hyman, Ray. "The Mischief Making of Ideomotor Action." *The Scientific Review of Alternative Medicine* 3, no. 2 (1999): 34–43.
Jacobson, Anne Jaap, ed. *Hubert Dreyfus and Problem of Representation*. Special Issue of *Phenomenology and the Cognitive Sciences* 1, no. 4, (2002).
James, William. "The Gospel of Relaxation." In *Writings 1878–1899*. Edited by Gerald E. Meyers, 825–40. New York: The Library of America, 1992.
Katz, David. "The Vibratory Sense." In *The Vibratory Sense and Other Lectures*, 90–103. Orono, ME: University Press, 1930.
Kauffman, Robert. "Tactility as an Aesthetic Consideration in African Music." In *The Performing Arts: Music and Dance*. Edited by John Blacking and Joann W. Kealiinohomoku, 251–53. The Hague: Mouton, 1979.

Kleeman, Janice E. "The Parameters of Musical Transmission." *The Journal of Musicology* 4 (1985/86): 1–22.

Kleist, Heinrich von. "Reflection: A Paradox." In *Selected Writings*. Edited and translated by David Constantine, 410. London: J. M. Dent, 1997.

Koffka, Kurt. *The Growth of the Mind: An Introduction to Child Psychology*. New Brunswick, NJ: Transaction Books, 1980.

Kott, Jan. "The Memory of the Body." In *The Memory of the Body: Essays on Theater and Death*, 113–19. Evanston: Northwestern University Press, 1992.

Kwant, Remigius C. *Phenomenology of Expression*. Atlantic Highlands, NJ: Humanities Press, 1978.

Ladrière, Jean. "La ville, inducteur existentiel." In *Vie sociale et destinée*, 139–60. Gembloux: Duculot, 1973.

Laing, R. D. *The Divided Self: An Existential Study in Sanity and Madness*. Harmondsworth: Penguin Books, 1965.

Lange, Roderyk. *The Nature of Dance: An Anthropological Perspective*. London: Macdonald & Evans, 1975.

Langer, Susanne K. *Philosophy in a New Key: A Study in the Symbolism of Reason, Rite, and Art*. 3rd ed. Cambridge, MA: Harvard University Press, 1957.

Lasch, Christopher. *The Culture of Narcissism: American Life in an Age of Diminishing Expectations*. New York: W. W. Norton, 1991.

Lavelle, Louis. *The Dilemma of Narcissus*. Translated by W. T. Gairdner. London: George Allen & Unwin, 1973.

Lecoq, Jacques, with Jean-Gabriel Carasso and Jean-Claude Lallias. *The Moving Body: Teaching Creative Theatre*. Translated by David Bradby. New York: Routledge, 2001.

Leder, Drew. *The Absent Body*. Chicago: University of Chicago Press, 1990.

Lennep, D. J. van. "The Psychology of Driving a Car." In *Phenomenological Psychology: The Dutch School*, edited by Joseph J. Kockelmans, 217–27. Dordrecht: Martinus Nijhoff, 1987.

Lersch, Philipp, *Aufbau der Person*. 11th ed. Munich: Johann Ambrosius Barth, 1970.

Lessschaeve, Jacqueline. *The Dancer and the Dance: Conversation with Merce Cunningham*. New York: Marion Boyars, 1999.

Lewis, C. S. *Studies in Words*. 2nd ed. Cambridge: Cambridge University Press, 1967.

Linschoten, Jan. "Aspects of the Sexual Incarnation. An Inquiry Concerning the Meaning of the Body in Sexual Encounter." In *Phenomenological Psychology: The Dutch School*, edited by Joseph J. Kockelmans, 149–94. Dordrecht: Martinus Nijhoff, 1987.

Lorenz, Konrad. *Behind the Mirror: A Search for a Natural History of Human Knowledge*. Translated by Ronald Taylor. New York: Harcourt Brace Jovanovich, 1977.

Maldiney, Henri. "Comprendre." *Revue de Métaphysique et de Morale* 1–2 (1961): 35–89.

———. *Penser l'homme et la folie: À la lumière de l'analyse existentielle et de l'analyse du destin*. Grenoble: Jérôme Millon, 1991.

Marcel, Gabriel. "Leibliche Begegnung: Notizen aus einem gemeinsamen Gedankengang." In *Leiblichkeit: Philosophische, gesellschaftliche und therapeutische Perspektiven*, edited by Hilarion Petzold, 15–46. Padeborn: Junfermann-Verlag, 1986.

Martin, John. *The Dance in Theory* (1939). E-book reprint. Highstown, NJ: Princeton Book Company, 2004.

Mazzù, Antonio. "Syntaxe motrice et stylistique corporelle: Réflexions à propos du schématisme corporel chez le premier Merleau-Ponty." *Revue philosophique de Louvain* 99 (2001): 46–72.

Mead, Margaret. "Balinese Character." In Gregory Bateson and Margaret Mead. *Balinese Character: A Photographic Analysis*, 1–48. New York: New York Academy of Sciences, 1942.

Meltzoff, Andrew N., and M. Keith Moore. "Infant's Understanding of People and Things: From Body Imitation to Folk Psychology." In *The Body and the Self*, edited by Jose Luis Bermudez, Anthony Marcel and Naomi Eilan, 43–69. Cambridge, MA: MIT Press, 1995.

Menuhin, Yehudi. "Improvisation and Interpretation." In *Theme and Variations*, 35–46. New York: Stein and Day, 1972.

Merleau-Ponty, Maurice. "Indirect Language and the Voices of Silence." In *Signs*. Translated by Richard C. McClearly, 39–83. Evanston: Northwestern University Press, 1964.

———. "Man and Adversity." In *Signs*. Translated by Richard C. McClearly, 224–43. Evanston: Northwestern University Press, 1964.

———. *The Structure of Behavior*. Translated by Alden L. Fisher. Boston: Beacon Press, 1967.

———. *Consciousness and Language Acquisition*. Translated by Hugh J. Silvermann. Evanston: Northwestern University Press, 1973.

———. *Phenomenology of Perception*. Translated by Colin Smith. London: Routledge, 1989.

Minkowski, Eugène. "Spontaneity (...Spontaneous Movement Like This!)." In *Readings in Existential Phenomenology*, edited by Nathaniel Lawrence and Daniel O'Connor, 168–77. Englewood: Prentice-Hall, 1967.

———. "Se répandre (L'olfactif)." In *Vers une cosmologie: Fragments philosophiques*, 111–20. Paris: Aubier-Montaigne, 1967.

———. "Voyons-nous avec les yeux ?." In *Vers une cosmologie: Fragments philosophiques*, 131–41. Paris: Aubier-Montaigne, 1967.

———. *Lived Time: Phenomenological and Psychopathological Studies*. Translated by Nancy Metzel. Evanston: Northwestern University Press, 1970.

Mooij, Anton. "Towards an Anthropological Psychiatry." *Theoretical Medicine*, 16 (1994): 73–91.

Mumford, Lewis. *Art and Technics*. New York: Columbia University Press, 1960.

Nagatomo, Shigenori. "Two Contemporary Japanese Views of the Body: Ichikawa Hiroshi and Yuasa Yasuo." In *Self as Body in Asian Theory and Practice*, edited by Thomas P. Kasulis, 321–46. Albany: State University of New York Press, 1993.

Nettl, Bruno. "Thoughts on Improvisation: A Comparative Approach." *The Musical Quarterly* 60 (1974): 1–19.

Nogué, Jean. *Esquisse d'un système des qualités sensibles*. Paris: PUF, 1943.

Nuland, Sherwin B. *The Wisdom of the Body*. New York: Alfred A. Knopf, 1997.

Ortega y Gasset, José. *Man and People*. Translated by Willard R. Trask. New York: W. W. Norton, 1963.

Olsen, Bjørnar. "Material Culture after Text: Re-Membering Things." *Norwegian Archeological Review* 36, no. 2 (2003): 87–104.

Palágyi, Menyhért (Melchior). *Székely Bertalan és a festészet aesthetikája* (Bertalan Székely and the Aesthetics of Painting). Budapest: Hoffmann & Vastagh, 1910.

Palágyi, Melchior. *Naturphilosophische Vorlesungen: Über die Grundprobleme des Bewusstseins und des Lebens*. Vol. 1 of *Ausgewählte Werke*. Leipzig: Johann Ambrosius Barth, 1924.

———. *Wahrnehmungslehre*. Vol. 2 of *Ausgewählte Werke*. Leipzig: Johann Ambrosius Barth, 1925.

Peursen, Cornelius A. van. *Body, Soul, Spirit: A Survey of the Body-Mind Problem*. Translated by Hubert H. Hoskins. London: Oxford University Press, 1966.

Piaget, Jean. *Play, Dreams and Imitation in Childhood*. Translated by C. Gattegno and F. M. Hodgson. New York: W. W. Norton, 1962.

Pike, Alfred. "A Phenomenology of Jazz." *Journal of Jazz Studies* 2 (1974): 88–94.

Plessner, Helmuth. *Laughing and Crying: A Study of the Limits of Human Behavior*. Translated by James S. Churchill and Marjorie Grene. Evanston: Northwestern University Press, 1970.

———. "Zur Anthropologie der Musik." In *Ausdruck und menschliche Natur*. Vol. 7 of *Gesammelte Schriften*, 184–201. Frankfurt am Main: Suhrkamp Verlag, 1982.

———. "Zur Anthropologie der Nachahmung." In *Ausdruck und menschliche Natur*. Vol. 7 of *Gesammelte Schriften*, 391–98. Frankfurt am Main: Suhrkamp Verlag, 1982.

———. "Der imitatorische Akt." In *Ausdruck und menschliche Natur.* Vol. 7 of *Gesammelte Schriften*, 449–57. Frankfurt am Main: Suhrkamp Verlag, 1982.

———. *The Limits of Community.* Translated by Andrew Wallace. Amherst, NY: Humanity Books, 1999.

Plügge, Herbert. *Wohlbefinden und Missbefinden: Beiträge zu einer medizinischen Anthropologie.* Tübingen: Max Niemeyer Verlag, 1962.

———. *Vom Spielraum des Leibes: Klinisch-phänomenologische Erwägungen über "Körperschema" und "Phantomglied".* Salzburg: Otto Müller Verlag, 1970.

Pohlen, Manfred. "Über die Beziehung zwischen rhythmischer Einstimmung und frühzeitiger Differenzierung des Gehörsinns bei der Entstehung des Ich und der Sprache." *Jahrbuch für Psychologie, Psychotherapie und medizinische Anthropologie* 17 (1969): 281–308.

Polanyi, Michael. *Personal Knowledge: Towards a Post-Critical Philosophy.* Chicago: University of Chicago Press, 1962.

———. *The Tacit Dimension.* New York: Anchor Books, 1967.

Pressing, Jeff. "Cognitive Processes in Improvisation." In *Cognitive Processes in the Perception of Art*, edited by Ray W. Crozier and Antony J. Chapman, 345–63. Amsterdam: Elsevier Science, 1984.

———. "Improvisation: Methods and Models." In *Generative Processes in Music: The Psychology of Performance, Improvisation, and Composition.* Edited by John A. Sloboda, 129–78. Oxford: Clarendon Press, 1988.

Prinz, Wolfgang. "Ideo-Motor Action." In *Perspectives on Perception and Action* Edited by Herbert Heuer and Andries F. Sanders, 47–76. Hillsdale, NJ: Lawrence Erlbaum Associates, 1987.

Prinzhorn, Hans. *Leib-Seele-Einheit: Ein Kernproblem der neuen Psychologie.* Potsdam: Müller & Kiepenheuer Verlag, 1927.

———. *Artistry of the Mentally Ill: A Contribution to the Psychology and Psychopathology of Configuration.* Translated by Eric von Brockdorff. New York: Springer-Verlag, 1962.

Révész, Géza. "La fonction sociologique de la main humaine et de la main animale." *Journal de psychologie normale et pathologique* 35 (1938): 26–49.

———. "Die Sprachfunktion der Hand." *Psychologische Beiträge* 2 (1955): 254–65.

Ricoeur, Paul. *Freedom and Nature: The Voluntary and the Involuntary.* Translated by Erazim V. Kohák. Evanston: Northwestern University Press, 1966.

Rivlin, Robert, and Karen Gravelle. *Deciphering the Senses: The Expanding World of Human Perception.* New York: Simon and Schuster, 1984.

Roseman, Janet Lynn. *Dance Masters: Interviews with Legends of Dance.* New York: Routledge, 2001.

Röthig, Peter. *Rhythmus und Bewegung: Eine Analyse aus der Sicht der Leibeserziehung.* 2nd ed. Schorndorf: Verlag Karl Hofmann, 1984.

———. "Bewegung – Rhythmus – Gestaltung: Zu Problemen gymnastischer Kategorien." In *Gymnastik: Ein Beitrag zur Bewegungskultur unserer Gesellschaft*, edited by Klaus-Jürgen Gutsche and Hans Jochen Medau, 36–51. Schorndorf: Verlag Karl Hofmann, 1989.

———. "Betrachtungen zur Körper- und Bewegungsästhetik." In *Grundlagen und Perspektiven ästhetischer und rhythmischer Bewegungserziehung*, edited by Eva Bannmüller and Peter Röthig, 85–97. Stuttgart: Ernst Klett Verlag, 1990.

———. "Zur Theorie des Rhythmus." In *Grundlagen und Perspektiven ästhetischer und rhythmischer Bewegungserziehung*, edited by Eva Bannmüller and Peter Röthig, 51–71. Stuttgart: Ernst Klett Verlag, 1990.

———. "Bewegungsgestaltung and ästhetische Erziehung im Sport." In *Facetten der Sportpädagogik*, edited by Robert Prohl, 13–22. Schorndorf: Verlag Karl Hofmann, 1993.

Rudert, J. "Die persönliche Atmosphäre." *Archiv für die gesamte Psychologie* 116 (1964): 291–98.

Rüssel, Arnulf. "Gestalt und Bewegung: Psychologische Grundfragen der Sprechspur." *Psychologische Beiträge* 2 (1955): 409–38.

Sacks, Oliver. *A Leg to Stand On*. New York: Harper Perennial, 1990.

Sándor, György. *On Piano Playing: Motion, Sound and Expression*. New York: Schirmer Books, 1995.

Sarath, Ed. "A New Look at Improvisation." *Journal of Music Theory* 40 (1996): 1–38.

Sarles, Harvey B. *Language and Human Nature*. Minneapolis: University of Minnesota Press, 1985.

Scheflen, Albert E. "Comments on the Significance of Interaction Rhythms." In *Interaction Rhythms: Periodicity in Communicative Behavior*, edited by Martha Davis, 13–22. New York: Human Sciences Press, 1982.

Scheler, Max. *On the Eternal in Man*. Translated by Bernard Noble, New York: Harper & Brothers, 1960.

Schmitz, Hermann. *Der Leib im Spiegel der Kunst*. Vol. II, 2 of *System der Philosophie*. Bonn: Bouvier Verlag, 1966.

———. *Leib und Gefühl: Materialen zu einer philosophischen Therapeutik*. Edited by Hermann Gausebeck and Gerhard Risch. Padeborn: Junfermann-Verlag, 1989.

Schutz, Alfred. "Making Music Together. A Study of Social Relationship." In *Studies in Social Theory*. Vol. 2 of *Collected Papers*. Edited by Maurice Natanson, 159–78. The Hague: Martinus Nijhoff, 1964.

Schwartz, Ulrich. "Walter Benjamin: Mimesis und Erfahrung." In *Philosophie der Gegenwart*. Vol. 6 of *Grundprobleme der grossen Philosophen*. Edited by Josef Speck, 43–77. Göttingen: Vandenhoeck und Ruprecht, 1984.

Seashore, Carl E. *Psychology of Music*. New York: Dover, 1967.
Seel, Martin. "Die Zelebration des Unvermögen: Zur Ästhetik des Sports." *Deutsche Zeitschrift für europäisches Denken* 47 (1993): 91–100.
Selle, Gert. *Gebrauch der Sinne: Eine kunstpädagogische Praxis*. Reinbek: Rowohlt Verlag, 1988.
Sennett, Richard. *Flesh and Stone: The Body and the City in Western Civilization*. New York: W. W. Norton, 1994.
———. "Resistance." *Granta 76: Music* (2002): 29–38.
Sheets-Johnstone, Maxine. *The Phenomenology of Dance*. New York: Books for Libraries, 1980.
Simmel, Georg. "Aesthetic Significance of the Face." In *Essays on Sociology, Philosophy and Aesthetics*. Edited by Kurt Wolff, 276–81. New York: Harper Torchbooks, 1959.
Singer, Linda. "Merleau-Ponty on the Concept of Style." *Man and World* 14 (1981): 153–63.
Sobotka, Raimund. *Formgesetze der Bewegungen im Sport*. Schorndorf: Verlag Karl Hofmann, 1974.
Souriau, Paul. *Aesthetics of Movement*. Translated and edited by Manon Souriau. Amherst: University of Massachusetts Press, 1983.
Spaemann, Robert. *Basic Moral Concepts*. Translated by T. J. Armstrong. London: Routledge, 1989.
Spiegelberg, Herbert. "On the Motility of the Ego: A Contribution to the Phenomenology of the Ego and Postscript 1978." In *Phenomenology: Critical Concepts in Philosophy*. Vol. 2. Edited by Dermot Moran and Lester E. Embree, 217–35. New York: Routledge, 2004.
Spitz, Hermann H. *Nonconscious Movements: From Mystical Messages to Facilitated Communication*. Manwah, NJ: Lawrence Erlbaum Associates, 1997.
Spitz, René A. *No and Yes: On the Genesis of Human Communication*. New York: International Universities Press, 1957.
———. *The First Year of Life: A Psychoanalytic Study of Normal and Deviant Development of Object Relations*. New York: International University Press, 1965.
Stern, Daniel N. *The Interpersonal World of the Infant: A View from Psychoanalysis and Developmental Psychology*. New York: Basic Books, 1985.
———. *The First Relationship: Infant and Mother*. Cambridge, MA: Harvard University Press, 2002.
Stewart, Susan. "Prologue: From the Museum of Touch." In *Material Memories: Design and Evocation*, edited by Marius Kwint, Christopher Breward and Jeremy Aynsley, 17–36. Oxford: Berg, 1999.
Storr, Anthony. *The Dynamics of Creation*. London: Secker & Warburg, 1972.
———. *Music and the Mind*. New York: Free Press, 1992.

Straus, Erwin W. *The Primary World of Senses: A Vindication of Sensory Experience.* Translated by Jacob Needleman. New York: Free Press of Glencoe, 1963.

———. "The Forms of Spatiality." In *Phenomenological Psychology* (1966). Reprint. Translated by Erling Eng, 3–37. New York: Garland, 1980.

———. *Man, Time, and World. Two Contributions to Anthropological Psychology.* Translated by Donald Moss. Pittsburgh: Duquesne University Press, 1982.

Sudnow, David. *Ways of the Hand: The Organization of Improvised Conduct.* Cambridge, MA: MIT Press, 1993.

Taylor, Charles. *The Ethics of Authenticity.* Cambridge, MA: Harvard University Press, 1992.

Taylor, Paul. "Down with Choreography." In *The Modern Dance: Seven Statements of Belief*, edited by Selma Jeanne Cohen, 91–102. Middletown, CT: Wesleyan University Press, 1969.

Tellenbach, Hubertus. *Geschmack und Atmosphäre: Medien menschlichen Elementarkontaktes.* Salzburg: Otto Müller Verlag, 1968.

———. "Die Begründung psychiatrischer Erfahrung und psychiatrischer Methoden in philosophischen Konzeptionen vom Wesen des Menschen." In *Philosophische Anthropologie: Erster Teil*, edited by Hans-Georg Gadamer and Paul Vogler, 138–81. Stuttgart: Georg Thieme Verlag; Munich: Deutscher Taschenbuch Verlag, 1974.

———. *Melancholy. History of the Problem, Endogeneity, Typology, Pathogenesis, Clinical Consideration.* Translated by Erling Eng. Pittsburgh: Duquesne University Press, 1980.

———. "The Education of Medical Student." In *The Moral Sense in the Communal Significance of Life.* Vol. 20 of *Analecta Husserliana.* Edited by Anna-Teresa Tymieniecka, 175–84. Dordrecht: D. Riedel, 1986.

Ten Have, Henk. "The Anthropological Tradition in the Philosophy of Medicine." *Theoretical Medicine*, 16 (1994): 3–14.

Valéry, Paul. "Degas, Dance, Drawing." In *Degas, Manet, Morisot.* Translated by David Paul, 5–102. New York: Pantheon Books, 1960.

———. "Philosophy of the Dance." In *Aesthetics.* Translated by Ralph Manheim, 197–211. New York: Pantheon Books, 1964.

———. *Cahiers.* Vol. 1. Paris: Gallimard, Pléiade, 1973.

Verdeau-Paillès, Jacqueline. "Music and the Body." In *The Fourth International Symposium on Music in Rehabilitation and Human Well-Being*, edited by Rosalie Rebollo Pratt, 37–48. Lanham, MD: University Press of America, 1987.

Vizinczey, Stephen. *An Innocent Millionaire.* Boston: The Atlantic Monthly Press, 1985.

Waldenfels, Bernhard. *Das leibliche Selbst: Vorlesungen zur Phänomenologie des Leibes.* Frankfurt am Main: Suhrkamp Verlag, 2000.

———. "Vom Rhythmus der Sinnen." In *Sinnesschwellen: Studien zur Phänomenologie des Fremden*, 79–83. Frankfurt am Main: Suhrkamp Verlag, 1999.

Walker, Alan. *Reflections on Liszt.* Ithaca, NY: Cornell University Press, 2005.

Wallin, Nils Lennart. *Biomusicology: Neurophysiological, Neuropsychological, and Evolutionary Perspectives on the Origins and Purposes of Music.* Stuyvesant, NY: Pendragon Press, 1991.

Weiss, Paul. "Man's Existence." *International Philosophical Quarterly* 1 (1961): 545–68.

———. *Privacy.* Carbondale: Southern Illinois University Press, 1983.

Werner, Heinz. *Comparative Psychology of Mental Development.* Rev. ed. New York: International Universities Press, 1980.

Whitehead, Alfred North. *The Aims of Education and Other Essays.* New York: Free Press, 1968.

———. *Modes of Thought.* New York: Free Press, 1968.

Wittgenstein, Ludwig. *Zettel.* Translated by G.E.M. Anscombe. Berkeley: University of California Press, 1970.

Yamaguchi, Ichiro. *Ki als leibhaftige Vernunft: Beitrag zur interkulturellen Phänomenologie der Leiblichkeit.* Munich: Wilhelm Fink Verlag, 1997.

Yudell, Robert J. "Body Movement." In Kent C. Bloomer and Charles W. Moore. *Body, Memory, and Architecture*, 57–76. New Haven: Yale University Press, 1977.

Zaner, Richard M. "The Discipline of the 'Norm:' A Critical Appreciation of Erwin Straus." *Human Studies*, 27 (2004): 37–50.

Zuboff, Shoshana. *In the Age of the Smart Machine: The Future of Work and Power.* New York: Basic Books, 1988.

Zutt, Jürg, "Die innere Haltung." In *Auf dem Wege zu einer anthropologischen Psychiatrie: Gesammelte Aufsätze*, 1–88. Berlin: Springer-Verlag, 1963.

———. "Der Leib der Tiere." In *Auf dem Wege zu einer anthropologischen Psychiatrie: Gesammelte Aufsätze*, 448–61. Berlin: Springer-Verlag, 1963.

———. "Über den tragenden Leib." In *Auf dem Wege zu einer anthropologischen Psychiatrie: Gesammelte Aufsätze*, 416–26. Berlin: Springer-Verlag, 1963.

index